"A candid and highly entertaining explanation of how and why a man who spent decades picking tech winners and cheering his industry on has been carried to the shore of social activism." —*The New York Times Book Review*

"A timely reckoning with Facebook's growth and data-obsessed culture . . . [*Zucked*] is the first narrative tale of Facebook's unravelling over the past two years. . . . McNamee excels at grounding Facebook in the historical context of the technology industry." —*Financial Times*

"[An] excellent new book . . . [McNamee] is one of the social network's biggest critics. He's a canny and persuasive one too. In *Zucked*, McNamee lays out an argument why it and other tech giants have grown into a monstrous threat to democracy. Better still he offers tangible solutions. . . . What makes McNamee so credible is his status as a Silicon Valley insider. He also has a knack for distilling often complex or meandering TED Talks and Medium posts about the ills of social media into something comprehensible, not least for those inside the D.C. Beltway. . . . McNamee doesn't just scream fire, though. He also provides a reasonable framework for solving some of the issues. . . . For anyone looking for a primer on what's wrong with social media and what to do about it, the book is well worth the read." —*Reuters*

"Think of *Zucked* as the story after *Social Network*'s credits roll. McNamee, an early Facebook investor and Zuckerberg mentor, weaves together a story of failed leadership, bad actors, and algorithms against the backdrop of the 2016 presidential election." —*Hollywood Reporter*

"McNamee's work is both a first-rate history of social media and a cautionary manifesto protesting their often overlooked and still growing dangers to human society." —*Booklist*

"Regardless of where you stand on the issue, you'll want to see why one of Facebook's biggest champions became one of its fiercest critics." —*Business Insider*

"A comprehensible primer on the political pitfalls of big tech." —*Publishers Weekly*

"Part memoir, part indictment, *Zucked* chronicles Facebook's history to demonstrate that its practices of 'invasive surveillance, careless sharing of private data, and behavior modification in pursuit of unprecedented scale and influence,' far from being a series of accidental oversights, were in fact foundational to the company's astronomical success. This historical approach allows McNamee to draw valuable connections between present-day troubles and the company's philosophical source code." —*Bookforum*

"Roger McNamee's *Zucked* fully captures the disastrous consequences that occur when people running companies wielding enormous power don't listen deeply to their stakeholders, fail to exercise their ethical responsibilities, and don't make trust their number one value."

—Marc Benioff, chariman and co-CEO of Salesforce

"McNamee puts his finger on serious problems in online environments, especially social networking platforms. I consider this book to be a must-read for anyone wanting to understand the societal impact of cyberspace."

—Vint Cerf, internet pioneer

"Roger McNamee is an investor with the nose of an investigator. This unafraid and unapologetic critique is enhanced by McNamee's personal association with Facebook's leaders and his long career in the industry. Whether you believe technology is the problem or the solution, one has no choice but to listen. It's only democracy at stake." —Emily Chang, author of *Brotopia*

"Roger McNamee is truly the most interesting man in the world—legendary investor, virtuoso guitarist, and damn lucid writer. He's written a terrific book that is both soulful memoir and muckraking exposé of social media. Everyone who spends their day staring into screens needs to read his impassioned tale."

—Franklin Foer, author of *World Without Mind*

"A frightening view behind the scenes of how absolute power and panoptic technologies can corrupt our politics and civic commons in this age of increasing-returns monopolies. Complementing Jaron Lanier's recent warnings with a clear-eyed view of politics, antitrust, and the law, this is essential reading for activists and policymakers as we work to preserve privacy and decency and a civil society in the internet age."

—Bill Joy, cofounder of Sun Microsystems, creator of the Berkeley Unix operating system

"*Zucked* is the mesmerizing and often hilarious story of how Facebook went from young darling to adolescent menace, not to mention a serious danger to democracy. With revelations on every page, you won't know whether to laugh or weep."

—Tim Wu, author of *The Attention Merchants* and *The Curse of Bigness*

"A well-reasoned and well-argued case against extractive technology."

—*Kirkus Reviews*

PENGUIN BOOKS

ZUCKED

Roger McNamee has been a Silicon Valley investor for thirty-five years. He cofounded successful funds in venture, crossover, and private equity. His most recent fund, Elevation, included U2's Bono as a cofounder. He holds a B.A. from Yale University and an M.B.A. from the Tuck School of Business at Dartmouth College. Roger plays bass and guitar in the bands Moonalice and Doobie Decibel System and is the author of *The New Normal* and *The Moonalice Legend: Posters and Words, Volumes 1–9*. He has served as a technical advisor for seasons two through six of HBO's "Silicon Valley" series and was also responsible for raising the money that created the Wikimedia Foundation.

ZUCKED

Waking Up to the Facebook Catastrophe 🔍

Roger McNamee

PENGUIN BOOKS

PENGUIN BOOKS

An imprint of Penguin Random House LLC

penguinrandomhouse.com

First published in the United States of America by Penguin Press,
an imprint of Penguin Random House LLC, 2019
This revised edition published in Penguin Books 2020

Copyright © 2019, 2020 by Roger McNamee
Penguin supports copyright. Copyright fuels creativity, encourages
diverse voices, promotes free speech, and creates a vibrant culture. Thank you
for buying an authorized edition of this book and for complying with copyright
laws by not reproducing, scanning, or distributing any part of it in any
form without permission. You are supporting writers and allowing
Penguin to continue to publish books for every reader.

"The Current Moment in History," remarks by George Soros delivered at the
World Economic Forum meeting, Davos, Switzerland, January 25, 2018.
Reprinted by permission of George Soros.

ISBN 9780525561378 (paperback)

THE LIBRARY OF CONGRESS HAS CATALOGED THE HARDCOVER EDITION AS FOLLOWS:
Names: McNamee, Roger, author.
Title: Zucked : waking up to the facebook catastrophe / Roger McNamee.
Description: New York : Penguin Press, 2019. | Includes bibliographical references and index.
Identifiers: LCCN 2018048578 (print) | LCCN 2018051479 (ebook) |
ISBN 9780525561361 (ebook) | ISBN 9780525561354 (hardcover) |
ISBN 9781984877895 (export)
Subjects: LCSH: Facebook (Electronic resource)—Social aspects. |
Online social networks—Political aspects—United States. |
Disinformation—United States. | Propaganda—Technological innovations. |
Zuckerberg, Mark, 1984– —Influence. | United States—Politics and government.
Classification: LCC HM743.F33 (ebook) | LCC HM743.F33 M347 2019 (print) |
DDC 302.30285—dc23
LC record available at https://lccn.loc.gov/2018048578

Printed in the United States of America
1   3   5   7   9   10   8   6   4   2

DESIGNED BY LUCIA BERNARD

While the author has made every effort to provide accurate telephone numbers, internet
addresses, and other contact information at the time of publication, neither the publisher nor
the author assumes any responsibility for errors or for changes that occur after publication.
Further, the publisher does not have any control over and does not assume any responsibility
for author or third-party websites or their content.

To Ann, who inspires me every day

Technology is neither good nor bad; nor is it neutral.

—*Melvin Kranzberg's First Law of Technology*

We cannot solve our problems with the same
thinking we used when we created them.

—*Albert Einstein*

Ultimately, what the tech industry really cares
about is ushering in the future, but it conflates
technological progress with societal progress.

—*Jenna Wortham*

CONTENTS

ZUCKED

Prologue

Technology is a useful servant but a
dangerous master. —Christian Lous Lange

November 9, 2016

"The Russians used Facebook to tip the election!"

So began my side of a conversation the day after the presidential election. I was speaking with Dan Rose, the head of media partnerships at Facebook. If Rose was taken aback by how furious I was, he hid it well.

Let me back up. I am a longtime tech investor and evangelist. Tech had been my career and my passion, but by 2016, I was backing away from full-time professional investing and contemplating retirement. I had been an early advisor to Facebook founder Mark Zuckerberg—Zuck, to many colleagues and friends—and an early investor in Facebook. I had been a true believer for a decade. Even at this writing, I still own shares in Facebook. In terms of my own narrow self-interest, I had no reason to bite Facebook's hand. It would never have occurred to me to be an anti-Facebook activist. I was more like Jimmy Stewart in

Hitchcock's *Rear Window*. He is minding his own business, checking out the view from his living room, when he sees what looks like a crime in progress, and then he has to ask himself what he should do. In my case, I had spent a career trying to draw smart conclusions from incomplete information, and one day early in 2016 I started to see things happening on Facebook that did not look right. I started pulling on that thread and uncovered a catastrophe. In the beginning, I assumed that Facebook was a victim and I just wanted to warn my friends. What I learned in the months that followed shocked and disappointed me. I learned that my trust in Facebook had been misplaced.

This book is the story of why I became convinced, in spite of myself, that even though Facebook provided a compelling experience for most of its users, it was terrible for America and needed to change or be changed, and what I have tried to do about it. My hope is that the narrative of my own conversion experience will help others understand the threat. Along the way, I will share what I know about the technology that enables internet platforms like Facebook to manipulate attention. I will explain how bad actors exploit the design of Facebook and other platforms to harm and even kill innocent people. How democracy has been undermined because of design choices and business decisions by internet platforms that deny responsibility for the consequences of their actions. How the culture of these companies causes employees to be indifferent to the negative side effects of their success. At this writing, there is nothing to prevent more of the same.

This is a story about trust. Technology platforms, including Facebook and Google, are the beneficiaries of trust and goodwill accumulated over fifty years by earlier generations of technology companies. They have taken advantage of our trust, using sophisticated techniques to prey on the weakest aspects of human psychology, to gather and exploit private data, and to craft business models that do not protect users from harm. Users must now learn to be skeptical about products they love, to change their online behavior, insist that platforms accept responsibility for the impact of their choices, and push policy makers to regulate the platforms to protect the public interest.

This is a story about privilege. It reveals how hypersuccessful people can be so focused on their own goals that they forget that others also have rights and privileges. How it is possible for otherwise brilliant people to lose sight of the fact that their users are entitled to self-determination. How success can breed overconfidence to the point of resistance to constructive feedback from friends, much less criticism. How some of the hardest working, most productive people on earth can be so blind to the consequences of their actions that they are willing to put democracy at risk to protect their privilege.

This is also a story about power. It describes how even the best of ideas, in the hands of people with good intentions, can still go terribly wrong. Imagine a stew of unregulated capitalism, addictive technology, and authoritarian values, combined with Silicon Valley's relentlessness and hubris, unleashed on billions of unsuspecting users. I think the day will come, sooner than I could have imagined just two years ago, when the world will recognize that the value users receive from the Facebook-dominated social media/attention economy revolution masked an unmitigated disaster for our democracy, for public health, for personal privacy, and for the economy. It did not have to be that way. It will take a concerted effort to fix it.

When historians finish with this corner of history, I suspect that they will cut Facebook some slack about the poor choices that Zuck, Sheryl Sandberg, and their team made as the company grew. I do. Making mistakes is part of life, and growing a startup to global scale is immensely challenging. Where I fault Facebook—and where I believe history will, as well—is for the company's response to criticism and evidence. They had an opportunity to be the hero in their own story by taking responsibility for their choices and the catastrophic outcomes those choices produced. Instead, Zuck and Sheryl chose another path.

This story is still unfolding. I have written this book now to serve as a warning. My goals are to make readers aware of a crisis, help them understand how and why it happened, and suggest a path forward. If I achieve only one thing, I hope it will be to make the reader appreciate that he or she has a role to play in the solution. I hope every reader will embrace the opportunity.

It is possible that the worst damage from Facebook and the other internet platforms is behind us, but that is not where the smart money will place its bet. The most likely case is that the technology and business model of Facebook and others will continue to undermine democracy, public health, privacy, and innovation until a countervailing power, in the form of government intervention or user protest, forces change.

TEN DAYS BEFORE the November 2016 election, I had reached out formally to Mark Zuckerberg and Facebook chief operating officer Sheryl Sandberg, two people I considered friends, to share my fear that bad actors were exploiting Facebook's architecture and business model to inflict harm on innocent people, and that the company was not living up to its potential as a force for good in society. In a two-page memo, I had cited a number of instances of harm, none actually committed by Facebook employees but all enabled by the company's algorithms, advertising model, automation, culture, and value system. I also cited examples of harm to employees and users that resulted from the company's culture and priorities. I have included the memo in the appendix.

Zuck created Facebook to bring the world together. What I did not know when I met him but would eventually discover was that his idealism was unbuffered by realism or empathy. He seems to have assumed that everyone would view and use Facebook the way he did, not imagining how easily the platform could be exploited to cause harm. He did not believe in data privacy and did everything he could to maximize disclosure and sharing. He operated the company as if every problem could be solved with more or better code. He embraced invasive surveillance, careless sharing of private data, and behavior modification in pursuit of unprecedented scale and influence. Surveillance, the sharing of user data, and behavioral modification are the foundation of Facebook's success. Users are fuel for Facebook's growth and, in some cases, the victims of it.

When I reached out to Zuck and Sheryl, all I had was a hypothesis that bad actors were using Facebook to cause harm. I suspected that the examples I saw reflected systemic flaws in the platform's design and the company's culture. I did not emphasize the threat to the presidential election, because at that time I could not imagine that the exploitation of Facebook would affect the outcome, and I did not want the company to dismiss my concerns if Hillary Clinton won, as was widely anticipated. I warned that Facebook needed to fix the flaws or risk its brand and the trust of users. While it had not inflicted harm directly, Facebook was being used as a weapon, and users had a right to expect the company to protect them.

The memo was a draft of an op-ed that I had written at the invitation of the technology blog *Recode*. My concerns had been building throughout 2016 and reached a peak with the news that the Russians were attempting to interfere in the presidential election. I was increasingly freaked out by what I had seen, and the tone of the op-ed reflected that. My wife, Ann, wisely encouraged me to send the op-ed to Zuck and Sheryl first, before publication. I had been one of Zuck's many advisors in Facebook's early days, and I played a role in Sheryl's joining the company as chief operating officer. I had not been involved with the company since 2009, but I remained a huge fan. My small contribution to the success of one of the greatest companies ever to come out of Silicon Valley was one of the true highlights of my thirty-four-year career. Ann pointed out that communicating through an op-ed might cause the wrong kind of press reaction, making it harder for Facebook to accept my concerns. My goal was to fix the problems at Facebook, not embarrass anyone. I did not imagine that Zuck and Sheryl had done anything wrong intentionally. It seemed more like a case of unintended consequences of well-intended strategies. Other than a handful of email exchanges, I had not spoken to Zuck in seven years, but I had interacted with Sheryl from time to time. At one point, I had provided them with significant value, so it was not crazy to imagine that they would take my concerns seriously. My goal was to persuade Zuck and Sheryl to investigate and take appropriate action. The publication of the op-ed could wait a few days.

Zuck and Sheryl each responded to my email within a matter of hours. Their replies were polite but not encouraging. They suggested that the problems I cited were anomalies that the company had already addressed, but they offered to connect me with a senior executive to hear me out. The man they chose was Dan Rose, a member of their inner circle with whom I was friendly. I spoke with Dan at least twice before the election. Each time, he listened patiently and repeated what Zuck and Sheryl had said, with one important addition: he asserted that Facebook was technically a platform, not a media company, which meant it was not responsible for the actions of third parties. He said it like that should have been enough to settle the matter.

Dan Rose is a very smart man, but he does not make policy at Facebook. That is Zuck's role. Dan's role is to carry out Zuck's orders. It would have been better to speak with Zuck, but that was not an option, so I took what I could get. Quite understandably, Facebook did not want me to go public with my concerns, and I thought that by keeping the conversation private, I was far more likely to persuade them to investigate the issues that concerned me. When I spoke to Dan the day after the election, it was obvious to me that he was not truly open to my perspective; he seemed to be treating the issue as a public relations problem. His job was to calm me down and make my concerns go away. He did not succeed at that, but he could claim one victory: I never published the op-ed. Ever the optimist, I hoped that if I persisted with private conversations, Facebook would eventually take the issue seriously.

I continued to call and email Dan, hoping to persuade Facebook to launch an internal investigation. At the time, Facebook had 1.7 billion active users. Facebook's success depended on user trust. If users decided that the company was responsible for the damage caused by third parties, no legal safe harbor would protect it from brand damage. The company was risking everything. I suggested that Facebook had a window of opportunity. It could follow the example of Johnson & Johnson when someone put poison in a few bottles of Tylenol on retail shelves in Chicago in 1982. J&J immediately withdrew every bottle of Tylenol from every retail location and did not reintroduce the product until it had perfected tamperproof packaging. The company absorbed a short-term hit to earnings but was rewarded with a huge increase in consumer trust. J&J had not put the poison in those bottles. It might have chosen to dismiss the problem as the work of a madman. Instead, it accepted responsibility for protecting its customers and took the safest possible course of action. I thought Facebook could convert a potential disaster into a victory by doing the same thing.

One problem I faced was that at this point I did not have data for making my case. What I had was a spidey sense, honed during a long career as a professional investor in technology.

I had first become seriously concerned about Facebook in February 2016, in the run-up to the first US presidential primary. As a political junkie, I was spending a few hours a day reading the news and also spending a fair amount of time on Facebook. I noticed a surge on Facebook of disturbing images, shared by friends, that originated on Facebook Groups ostensibly associated with the Bernie Sanders campaign. The images were deeply misogynistic depictions of Hillary Clinton. It was impossible for me to imagine that Bernie's campaign would allow them. More disturbing, the images were spreading virally. Lots of my friends were sharing them. And there were new images every day.

I knew a great deal about how messages spread on Facebook. For one thing, I have a second career as a musician in a band called Moonalice, and I had long been managing the band's Facebook page, which enjoyed high engagement with fans. The rapid spread of images from these Sanders-associated pages did not appear to be organic. How did the pages find my friends? How did my friends find the pages? Groups on Facebook do not emerge full grown overnight. I hypothesized that somebody had to be spending money on advertising to get the people I knew to join the Facebook Groups that were spreading the images. Who would do that? I had no answer. The flood of inappropriate images continued, and it gnawed at me.

More troubling phenomena caught my attention. In March 2016, for example, I saw a news report about a group that exploited a programming tool on Facebook to gather data on users expressing an interest in Black Lives Matter, data that they then sold to police departments, which struck me as evil. Facebook banned the group, but not until after irreparable harm had been done. Here again, a bad actor had used Facebook tools to harm innocent victims.

In June 2016, the United Kingdom voted to exit the European Union. The outcome of the Brexit vote came as a total shock. Polling had suggested that "Remain" would triumph over "Leave" by about four points, but precisely the opposite happened. No one could explain the huge swing. A possible explanation occurred to me. What if Leave had benefited from Facebook's architecture? The Remain campaign was expected to win because the UK had a sweet deal with the European Union: it enjoyed all the benefits of membership, while retaining its own currency. London was Europe's undisputed financial hub, and UK citizens could trade and travel freely across the open borders of the continent. Remain's "stay the course" message was based on smart economics but lacked emotion. Leave based its campaign on two intensely emotional appeals. It appealed to ethnic nationalism by blaming immigrants for the country's problems, both real and imaginary. It also promised that Brexit would generate huge savings that would be used to improve the National Health Service, an idea that allowed voters to put an altruistic shine on an otherwise xenophobic proposal.

The stunning outcome of Brexit triggered a hypothesis: in an election context, Facebook may confer advantages to campaign messages based on fear or anger over those based on neutral or positive emotions. It does this because Facebook's advertising business model depends on engagement, which can best be triggered through appeals to our most basic emotions. What I did not know at the time is that while joy also works, which is why puppy and cat videos and photos of babies are so popular, not everyone reacts the same way to happy content. Some people get jealous, for example. "Lizard brain" emotions such as fear and anger produce a more uniform reaction and are more viral in a mass audience. When users are riled up, they consume and share more content. Dispassionate users have relatively little value to Facebook, which does everything in its power to activate the lizard brain. Facebook has used surveillance to build giant profiles on every user and provides each user with a customized *Truman Show*, similar to the Jim Carrey film about a person who lives his entire life as the star of his own television show. It starts out giving users "what they want," but the algorithms are trained to nudge user attention in directions that Facebook wants. The algorithms choose posts calculated to press emotional buttons because scaring users or pissing them off increases time on site. When users pay attention, Facebook calls it *engagement*, but the goal is behavior modification that makes advertising more valuable. I wish I had understood this in 2016. At this writing, Facebook is the sixth most valuable company in America, despite being only fifteen years old, and its value stems from its mastery of surveillance and behavioral modification.

When new technology first comes into our lives, it surprises and astonishes us, like a magic trick. We give it a special place, treating it like the product equivalent of a new baby. The most successful tech products gradually integrate themselves into our lives. Before long, we forget what life was like before them. Most of us have that relationship today with smartphones and internet platforms like Facebook and Google. Their benefits are so obvious we can't imagine foregoing them. Not so obvious are the ways that technology products change us. The process has repeated itself in every generation since the telephone, including radio, television, and personal computers. On the plus side, technology has opened up the world, providing access to knowledge that was inaccessible in prior generations. It has enabled us to create and do remarkable things. But all that value has a cost. Beginning with television, technology has changed the way we engage with society, substituting passive consumption of content and ideas for civic engagement, digital communication for conversation. Subtly and persistently, it has contributed to our conversion from citizens to consumers. Being a citizen is an active state; being a consumer is passive. A transformation that crept along for fifty years accelerated dramatically with the introduction of internet platforms. We were prepared to enjoy the benefits but unprepared for the dark side. Unfortunately, the same can be said for the Silicon Valley leaders whose innovations made the transformation possible.

If you are a fan of democracy, as I am, this should scare you. Facebook has become a powerful source of news in most democratic countries. To a remarkable degree it has made itself the public square in which countries share ideas, form opinions, and debate issues outside the voting booth. But Facebook is more than just a forum. It is a profit-maximizing business controlled by one person. It is a massive artificial intelligence that influences every aspect of user activity, whether political or otherwise. Even the smallest decisions at Facebook reverberate through the public square the company has created with implications for every person it touches. The fact that users are not conscious of Facebook's influence magnifies the effect. If Facebook favors inflammatory campaigns, democracy suffers.

August 2016 brought a new wave of stunning revelations. Press reports confirmed that Russians had been behind the hacks of servers at the Democratic National Committee (DNC) and Democratic Congressional Campaign Committee (DCCC). Emails stolen in the DNC hack were distributed by WikiLeaks, causing significant damage to the Clinton campaign. The chairman of the DCCC pleaded with Republicans not to use the stolen data in congressional campaigns. I wondered if it were possible that Russians had played a role in the Facebook issues that had been troubling me earlier.

Just before I wrote the op-ed, ProPublica revealed that Facebook's advertising tools enabled property owners to discriminate based on race, in violation of the Fair Housing Act. The Department of Housing and Urban Development opened an investigation that was later closed, but reopened in April 2018. Here again, Facebook's architecture and business model enabled bad actors to harm innocent people.

Like Jimmy Stewart in the movie, I did not have enough data or insight to understand everything I had seen, so I sought to learn more. As I did so, in the days and weeks after the election, Dan Rose exhibited incredible patience with me. He encouraged me to send more examples of harm, which I did. Nothing changed. Dan never budged. In February 2017, more than three months after the election, I finally concluded that I would not succeed in convincing Dan and his colleagues; I needed a different strategy. Facebook remained a clear and present danger to democracy. The very same tools that made Facebook a compelling platform for advertisers could also be exploited to inflict harm. Facebook was getting more powerful by the day. Its artificial intelligence engine learned more about every user. Its algorithms got better at pressing users' emotional buttons. Its tools for advertisers improved constantly. In the wrong hands, Facebook was an ever-more-powerful weapon. And the next US election—the 2018 midterms—was fast approaching.

Yet no one in power seemed to recognize the threat. The early months of 2017 revealed extensive relationships between officials of the Trump campaign and people associated with the Russian government. Details emerged about a June 2016 meeting in Trump Tower between inner-circle members of the campaign and Russians suspected of intelligence affiliations. Congress spun up Intelligence Committee investigations that focused on that meeting.

But still there was no official concern about the role that social media platforms, especially Facebook, had played in the 2016 election. Every day that passed without an investigation increased the likelihood that the interference would continue. If someone did not act quickly, our democratic processes could be overwhelmed by outside forces; the 2018 midterm election would likely be subject to interference, possibly greater than we had seen in 2016. Our Constitution anticipated many problems, but not the possibility that a foreign country could interfere in our elections without consequences. I could not sit back and watch. I needed some help, and I needed a plan, not necessarily in that order.

1

The Strangest Meeting Ever

New technology is not good or evil in and of itself. It's all about how people choose to use it. —DAVID WONG

I should probably tell the story of how I intersected with Facebook in the first place. In the middle of 2006, Facebook's chief privacy officer, Chris Kelly, sent me an email stating that his boss was facing an existential crisis and required advice from an unbiased person. Would I be willing to meet with Mark Zuckerberg?

Facebook was two years old, Zuck was twenty-two, and I was fifty. The platform was limited to college students, graduates with an alumni email address, and high school students. News Feed, the heart of Facebook's user experience, was not yet available. The company had only nine million dollars in revenue in the prior year. But Facebook had huge potential—that was already obvious—and I leapt at the opportunity to meet its founder.

Zuck showed up at my Elevation Partners office on Sand Hill Road in Menlo Park, California, dressed casually, with a messenger bag over his shoulder. U2 singer Bono and I had formed Elevation in 2004, along with former Apple CFO Fred Anderson, former Electronic Arts president John Riccitiello, and two career investors, Bret Pearlman and Marc Bodnick. We had configured one of our conference rooms as a living room, complete with a large arcade video game system, and that is where Zuck and I met. We closed the door and sat down on comfy chairs about three feet apart. No one else was in the room.

Since this was our first meeting, I wanted to say something before Zuck told me about the existential crisis.

"If it has not already happened, Mark, either Microsoft or Yahoo is going to offer one billion dollars for Facebook. Your parents, your board of directors, your management team, and your employees are going to tell you to take the offer. They will tell you that with your share of the proceeds—six hundred and fifty million dollars—you will be able to change the world. Your lead venture investor will promise to back your next company so that you can do it again.

"It's your company, but I don't think you should sell. A big company will screw up Facebook. I believe you are building the most important company since Google and that before long you will be bigger than Google is today. You have two huge advantages over previous social media platforms: you insist on real identity and give consumers control over their privacy settings.

"In the long run, I believe Facebook will be far more valuable to parents and grandparents than to college students and recent grads. People who don't have much time will love Facebook, especially when families have the opportunity to share photos of kids and grandkids.

"Your board of directors, management team, and employees signed up for your vision. If you still believe in your vision, you need to keep Facebook independent. Everyone will eventually be glad you did."

This little speech took about two minutes to deliver. What followed was the longest silence I have ever endured in a one-on-one meeting. It probably lasted four or five minutes, but it seemed like forever. Zuck was lost in thought, pantomiming a range of *Thinker* poses. I have never seen anything like it before or since. It was painful. I felt my fingers involuntarily digging into the upholstered arms of my chair, knuckles white, tension rising to a boiling point. At the three-minute mark, I was ready to scream. Zuck paid me no mind. I imagined thought bubbles over his head, with reams of text rolling past. How long would he go on like this? He was obviously trying to decide if he could trust me. How long would it take? How long could I sit there?

Eventually, Zuck relaxed and looked at me. He said, "You won't believe this."

I replied, "Try me."

"One of the two companies you mentioned wants to buy Facebook for one billion dollars. Pretty much everyone has reacted the way you predicted. They think I should take the deal. How did you know?"

"I didn't know. But after twenty-four years, I know how Silicon Valley works. I know your lead venture investor. I know Yahoo and Microsoft. This is how things go around here."

I continued, "Do you want to sell the company?"

He replied, "I don't want to disappoint everyone."

"I understand, but that is not the issue. Everyone signed up to follow your vision for Facebook. If you believe in your vision, you need to keep Facebook independent. Yahoo and Microsoft will wreck it. They won't mean to, but that is what will happen. What do you want to do?"

"I want to stay independent."

I asked Zuck to explain Facebook's shareholder voting rules. It turned out he had a "golden vote," which meant that the company would always do whatever he decided. It took only a couple of minutes to figure that out. The entire meeting took no more than half an hour.

Zuck left my office and soon thereafter told Yahoo that Facebook was not for sale. There would be other offers for Facebook, including a second offer from Yahoo, and he would turn them down, too.

So began a mentorship that lasted three years. In a success story with at least a thousand fathers, I played a tiny role, but I contributed on two occasions that mattered to Facebook's early success: the Yahoo deal and the hiring of Sheryl. Zuck had other mentors, but he called on me when he thought I could help, which happened often enough that for a few years I was a regular visitor to Facebook's headquarters. Ours was a purely business relationship. Zuck was so amazingly talented at such a young age, and he leveraged me effectively. It began when Facebook was a little startup with big dreams and boundless energy. Zuck had an idealistic vision of connecting people and bringing them together. The vision inspired me, but the magic was Zuck himself. Obviously brilliant, Zuck possessed a range of characteristics that distinguished him from the typical Silicon Valley entrepreneur: a desire to learn, a willingness to listen, and, above all, a quiet confidence. Many tech founders swagger through life, but the best ones—including the founders of Google and Amazon—are reserved, thoughtful, serious. To me, Facebook seemed like the Next Big Thing that would make the world better through technology. I could see a clear path to one hundred million users, which would have been a giant success. It never occurred to me that success would lead to anything but happiness.

The only skin in the game for me at that time was emotional. I had been a Silicon Valley insider for more than twenty years. My fingerprints were on dozens of great companies, and I hoped that one day Facebook would be another. For me, it was a no-brainer. I did not realize then that the technology of Silicon Valley had evolved into uncharted territory, that I should no longer take for granted that it would always make the world a better place. I am pretty certain that Zuck was in the same boat; I had no doubt then of Zuck's idealism.

Silicon Valley had had its share of bad people, but the limits of the technology itself had generally prevented widespread damage. Facebook came along at a time when it was possible for the first time to create tech businesses so influential that no country would be immune to their influence. No one I knew ever considered that success could have a downside. From its earliest days, Facebook was a company of people with good intentions. In the years I knew them best, the Facebook team focused on attracting the largest possible audience, not on monetization. Persuasive technology and manipulation never came up. It was all babies and puppies and sharing with friends.

I am not certain when Facebook first applied persuasive technology to its design, but I can imagine that the decision was not controversial. Advertisers and media companies had been using similar techniques for decades. Despite complaints about television from educators and psychologists, few people objected strenuously to the persuasive techniques employed by networks and advertisers. Policy makers and the public viewed them as legitimate business tools. On PCs, those tools were no more harmful than on television. Then came smartphones, which changed everything. User count and usage exploded, as did the impact of persuasive technologies, enabling widespread addiction. That is when Facebook ran afoul of the law of unintended consequences. Zuck and his team did not anticipate that the design choices that made Facebook so compelling for users would also enable a wide range of undesirable behaviors. When those behaviors became obvious after the 2016 presidential election, Facebook first denied their existence, then responsibility for them. Perhaps it was a reflexive corporate reaction. In any case, Zuck, Sheryl, the team at Facebook, and the board of directors missed an opportunity to build a new trust with users and policy makers. Those of us who had advised Zuck and profited from Facebook's success also bear some responsibility for what later transpired. We suffered from a failure of imagination. The notion that massive success by a tech startup could undermine society and democracy did not occur to me or, so far as I know, to anyone in our community. Now the whole world is paying for it.

In the second year of our relationship, Zuck gave Elevation an opportunity to invest. I pitched the idea to my partners, emphasizing my hope that Facebook would become a company in Google's class. The challenge was that Zuck's offer would have us invest in Facebook *indirectly*, through a complicated, virtual security. Three of our partners were uncomfortable with the structure of the investment for Elevation, but they encouraged the rest of us to make personal investments. So Bono, Marc Bodnick, and I invested. Two years later, an opportunity arose for Elevation to buy stock in Facebook, and my partners jumped on it.

WHEN CHRIS KELLY CONTACTED ME, he knew me only by reputation. I had been investing in technology since the summer of 1982. Let me share a little bit of my own history for context, to explain where my mind was when I first entered Zuck's orbit.

I grew up in Albany, New York, the second youngest in a large and loving family. My parents had six children of their own and adopted three of my first cousins after their parents had a health crisis. One of my sisters died suddenly at two and a half while I was in the womb, an event that had a profound impact on my mother. At age two, I developed a very serious digestive disorder, and doctors told my parents I could not eat grains of any kind. I eventually grew out of it, but until I was ten, I could not eat a cookie, cake, or piece of bread without a terrible reaction. It required self-discipline, which turned out to be great preparation for the life I chose.

My parents were very active in politics and civil rights. The people they taught me to look up to were Franklin Roosevelt and Jackie Robinson. They put me to work on my first political campaign at age four, handing out leaflets for JFK. My father was the president of the Urban League in our home town, which was a big deal in the mid-sixties, when President Johnson pushed the Civil Rights Act and Voting Rights Act through Congress. My mother took me to a civil rights meeting around the time I turned nine so that I could meet my hero, Jackie Robinson.

The year that I turned ten, my parents sent me to summer camp.

During the final week, I had a terrible fall during a scavenger hunt. The camp people put me in the infirmary, but I was unable to keep down any food or water for three days, after which I had a raging fever. They took me to a nearby community hospital, where a former military surgeon performed an emergency operation that saved my life. My intestine had been totally blocked by a blood clot. It took six months to recover, costing me half of fourth grade. This turned out to have a profound impact on me. Surviving a near-death experience gave me courage. The recovery reinforced my ability to be happy outside the mainstream. Both characteristics proved valuable in the investment business.

My father worked incredibly hard to support our large family, and he did so well. We lived an upper-middle-class life, but my parents had to watch every penny. My older siblings went off to college when I was in elementary school, so finances were tight some of those years. Being the second youngest in a huge family, I was most comfortable observing the big kids. Health issues reinforced my quiet, observant nature. My mother used me as her personal Find My iPhone whenever she mislaid her glasses, keys, or anything. For some reason, I always knew where everything was.

I was not an ambitious child. Team sports did not play much of a role in my life. It was the sixties, so I immersed myself in the anti-war and civil rights movements from about age twelve. I took piano lessons and sang in a church choir, but my passion for music did not begin until I took up the guitar in my late teens. My parents encouraged me but never pushed. They were role models who prioritized education and good citizenship, but they did not interfere. They expected my siblings and me to make good choices. Through my teenage years, I approached everything but politics with caution, which could easily be confused with reluctance. If you had met me then, you might well have concluded that I would never get around to doing anything.

My high school years were challenging in a different way. I was a good student, but not a great one. I liked school, but my interests were totally different from my classmates'. Instead of sports, I devoted my free time to politics. The Vietnam War remained the biggest issue in the country, and one of my older brothers had already been drafted into the army. It seemed possible that I would reach draft age before the war ended. As I saw it, the rational thing to do was to work to end the war. I volunteered for the McGovern for President campaign in October 1971 and was in the campaign office in either New Hampshire or upstate New York nearly every day from October 1971, the beginning of my tenth-grade year, through the general election thirteen months later. That was the period when I fell in love with the hippie music of San Francisco: the Grateful Dead, Jefferson Airplane, Quicksilver Messenger Service, Big Brother and the Holding Company, and Santana.

I did not like my school, so once the McGovern campaign ended, I applied to School Year Abroad in Rennes, France, for my senior year. It was an amazing experience. Not only did I become fluent in French, I went to school with a group of people who were more like me than any set of classmates before them. The experience transformed me. I applied to Yale University and, to my astonishment, got in.

After my freshman year at Yale, I was awarded an internship with my local congressman, who offered me a permanent job as his legislative assistant a few weeks later. The promotion came with an increase in pay and all the benefits of a full-time job. I said no—I thought the congressman was crazy to promote me at nineteen—but I really liked him and returned for two more summers.

A year later, in the summer of 1976, I took a year off to go to San Francisco with my girlfriend. In my dreams, I was going to the city of the Summer of Love. By the time I got there, though, it was the city of Dirty Harry, more noir than flower power. Almost immediately, my father was diagnosed with inoperable prostate cancer. Trained as a lawyer, my father had started a brokerage firm that grew to a dozen offices. It was an undersized company in an industry that was undergoing massive change. He died in the fall of 1977, at a particularly difficult time for his business, leaving my mother with a house and little else. There was no money for me to return to college. I was on my own, with no college degree. I had my guitar, though, and practiced for many hours every day.

When I first arrived in San Francisco, I had four hundred dollars in my pocket. My dream of being a reporter in the mold of Woodward and Bernstein lasted for about half a day. Three phone calls were all it took to discover that there were no reporter jobs available for a college dropout like me, but every paper needed people in advertising sales. I was way too introverted for traditional sales, but that did not stop me. I discovered a biweekly French-language newspaper where I would be the entire advertising department, which meant not only selling ads but also collecting receivables from advertisers. When you only get paid based on what you collect, you learn to judge the people you sell to. If the ads didn't work, they wouldn't pay. I discovered that by focusing on multi-issue advertising commitments from big accounts, such as car dealerships, airlines, and the phone company, I could leverage my time and earn a lot more money per issue. I had no social life, but I started to build savings. In the two and a half years I was in San Francisco, I earned enough money to go back to Yale, which cost no more than 10 percent of what it costs today.

Every weekday morning in San Francisco I watched a locally produced stock market show hosted by Stuart Varney, who went on to a long career in broadcasting at CNN and Fox Business Network. After watching the show for six months and reading *Barron's* and stacks of annual reports, I finally summoned the courage to buy one hundred shares of Beech Aircraft. It went up 30 percent in the first week. I was hooked. I discovered that investing was a game, like Monopoly, but with real money. The battle of wits appealed to me. I never imagined then that investing would be my career. In the fall of 1978, I reapplied to Yale. They accepted me again, just weeks before two heartbreaking events chased me from San Francisco: the mass suicide of hundreds of San Franciscans at Jonestown and the murder of San Francisco's mayor and supervisor Harvey Milk by another member of the city's board of supervisors.

Celebrating my first Christmas at home since 1975, I received a gift that would change my life. My older brother George, ten years my senior, gave me a Texas Instruments Speak & Spell. Introduced just months earlier, the Speak & Spell combined a keyboard, a one-line alphanumeric display, a voice processor, and some memory to teach elementary school children to pronounce and spell words. But to my brother, it was the future of computing. "This means that in a few years, it will be possible to create a handheld device that holds all your personal information," he said.

He told me this in 1978. The Apple II had been introduced only a year earlier. The IBM PC was nearly three years in the future. The PalmPilot was more than eighteen years away. But my brother saw the future, and I took it to heart. I went back to college as a history major but was determined to take enough electrical engineering courses that I could design the first personal organizer. I soon discovered that electrical engineering requires calculus, and I had never taken calculus. I persuaded the professor to let me take the entry-level course anyway. He said if I did everything right except the math, he would give me a B ("for bravery"). I accepted. He tutored me every week. I took a second, easier engineering survey course, in which I learned concepts related to acoustics and mechanical engineering. I got catalogues and manuals and tried to design an oversized proof of concept. I could not make it work.

A real highlight of my second swing through Yale was playing in a band called Guff. Three guys in my dorm had started the band, but they needed a guitar player. Guff wrote its own songs and occupied a musical space somewhere near the intersection of the Grateful Dead, Frank Zappa, and punk rock. We played a ton of gigs, but college ended before the band was sufficiently established to justify making a career of it.

The band got paid a little money, but I needed to earn tuition-scale money. Selling ads paid far better than most student jobs, so I persuaded the Yale Law School Film Society to let me create a magazine-style program for their film series. I created a program for both semesters of senior year and earned almost enough money to pay for a year of graduate school.

But before that, in the fall of my senior year, I enrolled in Introduction to Music Theory, a brutal two-semester course for music majors. I was convinced that a basic knowledge of music theory would enable me to write better songs for my band. They randomly assigned me to one of a dozen sections, each with fifteen students, all taught by graduate students. The first class session was the best hour of classroom time I had ever experienced, so I told my roommate to switch from his section to mine. Apparently many others did the same thing, as forty people showed up the second day. That class was my favorite at Yale. The grad student who taught the class, Ann Kosakowski, did not teach the second semester, but early in the new semester, I ran into her as she exited the gymnasium, across the street from my dorm. She was disappointed because she had narrowly lost a squash match in the fifth game to the chair of the music department, so I volunteered to play her the next day. We played squash three days in a row, and I did not win a single point. Not one. But it didn't matter. I had never played squash and did not care about the score. Ann was amazing. I wanted to get to know her. I invited her on a date to see the Jerry Garcia Band right after Valentine's Day. A PhD candidate in music theory, Ann asked, "What instrument does Mr. Garcia play?" thinking perhaps it might be the cello. Ann and I are about to celebrate the fortieth anniversary of that first date.

Ann and I graduated together, she a very young PhD, me an old undergraduate. She received a coveted tenure-track position at Swarthmore College, outside of Philadelphia. I could not find a job in Philadelphia, so I enrolled at the Tuck School of Business at Dartmouth, in Hanover, New Hampshire. So began a twenty-one-year interstate commute.

My first job after business school was at T. Rowe Price, in Baltimore, Maryland. It was a lot closer to Philadelphia than Hanover, but still too far to commute every day. That's when I got hit by two game-changing pieces of good luck: my start date and my coverage group. My career began on the first day of the bull market of 1982, and they asked me to analyze technology stocks. In those days, there were no tech-only funds. T. Rowe Price was the leader in the emerging growth category of mutual funds, which meant they focused on technology more than anyone. I might not be able to make the first personal organizer, I reasoned, but I would be able to invest in it when it came along.

In investing, they say that timing is everything. By assigning me to cover tech on the first day of an epic bull market, T. Rowe Price basically put me in a position where I had a tailwind for my entire career. I can't be certain that every good thing in my career resulted from that starting condition, but I can't rule it out either. It was a bull market, so most stocks were going up. In the early days, I just had to produce reports that gave the portfolio managers confidence in my judgment. I did not have a standard pedigree for an analyst, so I decided to see if I could adapt the job to leverage my strengths.

I became an analyst by training, a nerd who gets paid to understand the technology industry. When my career started, most analysts focused primarily on financial statements, but I changed the formula. I have been successful due to an ability to understand products, financial statements, and trends, as well as to judge people. I think of it as real-time anthropology, the study of how humans and technology evolve and interact. I spend most of my time trying to understand the present so I can imagine what might happen in the future. From any position on the chessboard, there are only a limited number of moves. If you understand that in advance and study the possibilities, you will be better prepared to make good choices each time something happens. Despite what people tell you, the technology world does not actually change that much. It follows relatively predictable patterns. Major waves of technology last at least a decade, so the important thing is to recognize when an old cycle is ending and when a new one is starting. As my partner John Powell likes to say, sometimes you can see which body is tied to the railroad tracks before you can see who is driving the train.

The personal computer business started to take off in 1985, and I noticed two things: everyone was my age, and they convened at least monthly in a different city for a conference or trade show. I persuaded my boss to let me join the caravan. Almost immediately I had a stroke of good luck. I was at a conference in Florida when I noticed two guys unloading guitars and amps from the back of a Ford Taurus. Since all guests at the hotel were part of the conference, I asked if there was a jam session I could join. There was. It turns out that the leaders of the PC industry didn't go out to bars. They rented instruments and played music. When I got to my first jam session, I discovered I had an indispensable skill. Thanks to many years of gigs in bands and bars, I knew a couple hundred songs from beginning to end. No one else knew more than a handful. This really mattered because the other players included the CEO of a major software company, the head of R&D from Apple, and several other industry big shots. Microsoft cofounder Paul Allen played with us from time to time, but only on songs written by Jimi Hendrix. He could shred. Suddenly, I was part of the industry's social fabric. It is hard to imagine this happening in any other industry, but I was carving my own path.

My next key innovation related to earnings models. Traditional analysts used spreadsheets to forecast earnings, but spreadsheets tend to smooth everything. In tech, where success is binary, hot products always beat the forecast, and products that are not hot always fall short. I didn't need to worry about earnings models. I just needed to figure out which products were going to be hot. Forecasting products was not easy, but I did not need to be perfect. As with the two guys being chased by a bear, I only needed to do it better than the other guy.

I got my first chance to manage a portfolio in late 1985. I was asked to run the technology sector of one of the firm's flagship funds; tech represented about 40 percent of the fund. It was the largest tech portfolio in the country at the time, so it was a big promotion and an amazing opportunity. I had been watching portfolio managers for three years, but that did not really prepare me. Portfolio management is a game played with real money. Everyone makes mistakes. What differentiates great portfolio managers is their ability to recognize mistakes early and correct them. Portfolio managers learn by trial and error, with lots of errors. The key is to have more money invested in your good ideas than your bad ones.

T. Rowe launched a pure-play Science & Technology Fund, managed by two of my peers, on September 30, 1987. Nineteen days later, the stock market crashed. Every mutual fund got crushed, and Science & Tech was down 31 percent after only a month in business. While the number was terrible, it was actually better than competitors because the portfolio managers had invested only half their capital when the market collapsed. In the middle of 1988, with the viability of the fund in doubt, the firm reassigned the two managers and asked me to take over. I agreed to do so on one condition: I would run the fund my way. I told my bosses that I intended to be aggressive.

Another piece of amazing luck hit me when T. Rowe Price decided to create a growth-stage venture fund. I was already paying attention to private companies, because in those days, the competition in tech came from startups, not established companies. Over the next few years, I led three key growth-stage venture investments: Electronic Arts, Sybase, and Radius. The lead venture investor in all three companies was Kleiner Perkins Caufield & Byers, one of the leading venture capital firms in Silicon Valley. All three went public relatively quickly, making me popular both at T. Rowe Price and Kleiner Perkins. My primary contact at Kleiner Perkins was a young venture capitalist named John Doerr, whose biggest successes to that point had been Sun Microsystems,

Compaq Computer, and Lotus Development. Later, John would be the lead investor in Netscape, Amazon, and Google.

My strategy with the Science & Technology Fund was to focus entirely on emerging companies in the personal computer, semiconductor, and database software industries. I ignored all the established companies, a decision that gave the fund a gigantic advantage. From its launch through the middle of 1991, a period that included the 1987 crash and a second mini-crash in the summer of 1990, the fund achieved a 17 percent per annum return, against 9 percent for the S&P 500 and 6 percent for the technology index. That was when I left T. Rowe Price with John Powell to launch Integral Capital Partners, the first institutional fund to combine public market investments with growth-stage venture capital. We created the fund in partnership with Kleiner Perkins—with John Doerr as our venture capitalist—and Morgan Stanley. Our investors were the people who know us best, the founders and executives of the leading tech companies of that era.

Integral had a charmed run. Being inside the offices of Kleiner Perkins during the nineties meant we were at ground zero for the internet revolution. I was there the day that Marc Andreessen made his presentation for the company that became Netscape, when Jeff Bezos did the same for Amazon, and when Larry Page and Sergey Brin pitched Google. I did not imagine then how big the internet would become, but it did not take long to grasp its transformational nature. The internet would democratize access to information, with benefits to all. Idealism ruled. In 1997, Martha Stewart came in with her home-decorating business, which, thanks to an investment by Kleiner Perkins, soon went public as an internet stock, which seemed insane to me. I was convinced that a mania had begun for dot-coms, embodied in the Pets.com sock puppet and the slapping of a little "e" on the front of a company's name or a ".com" at the end. I knew that when the bubble burst, there would be a crash that would kill Integral if we did not do something radical.

I took my concerns to our other partner, Morgan Stanley, and they gave me some money to figure out the Next Big Thing in tech investing, a fund that could survive a bear market. It took two years, but Integral launched Silver Lake Partners, the first private equity fund focused on technology. Our investors shared our concerns and committed one billion dollars to the new fund.

Silver Lake planned to invest in mature technology companies. Once a tech company matured in those days, it became vulnerable to competition from startups. Mature companies tend to focus on the needs of their existing customers, which often blinds them to new business opportunities or new technologies. In addition, as growth slows, so too does the opportunity for employees to benefit from stock options, which startups exploit to recruit the best and brightest from established companies. My vision for Silver Lake was to reenergize mature companies by recapitalizing them to enable investment in new opportunities, while also replicating the stock compensation opportunities of a startup. The first Silver Lake fund had extraordinary results, thanks to three investments: Seagate Technology, Datek, and Gartner Group.

During the Silver Lake years, I got a call from the business manager of the Grateful Dead, asking for help. The band's leader, Jerry Garcia, had died a few years before, leaving the band with no tour to support a staff of roughly sixty people. Luckily, one of the band's roadies had created a website and sold merchandise directly to fans. The site had become a huge success, and by the time I showed up, it was generating almost as much profit as the band had made in its touring days. Unfortunately, the technology was out of date, but there was an opportunity to upgrade the site, federate it to other bands, and prosper as never before. One of the bands that showed an interest was U2. They found me through a friend of Bono's at the Department of the Treasury, a woman named Sheryl Sandberg. I met Bono and the Edge at Morgan Stanley's offices in Los Angeles on the morning after the band had won a Grammy for the song "Beautiful Day." I could not have named a U2 song, but I was blown away by the intelligence and business sophistication of the two Irishmen. They invited me to Dublin to meet their management. I made two trips during the spring of 2001.

On my way home from that second trip, I suffered a stroke. I didn't realize it at the time, and I tried to soldier on. Shortly thereafter, after some more disturbing symptoms, I found myself at the Mayo Clinic, where I learned that I had in fact suffered two ischemic strokes, in addition to something called a transient ischemic attack in my brain stem. It was a miracle I had survived the strokes and suffered no permanent impairment.

The diagnosis came as a huge shock. I had a reasonably good diet, a vigorous exercise regime, and a good metabolism, yet I had had two strokes. It turned out that I had a birth defect in my heart, a "patent foramen ovale," basically the mother of all heart murmurs. I had two choices: I could take large doses of blood thinner and live a quiet life, or I could have open-heart surgery and eliminate the risk forever. I chose surgery.

I had successful surgery in early July 2001, but my recovery was very slow. It took me nearly a year to recover fully. During that time, Apple shipped the first iPod. I thought it was a sign of good things to come and reached out to Steve Jobs to see if he would be interested in recapitalizing Apple. At the time, Apple's share price was about twelve dollars per share, which, thanks to stock splits, is equivalent to a bit more than one dollar per share today. The company had more than twelve dollars in cash per share, which meant investors were attributing zero value to Apple's business. Most of the management options had been issued at forty dollars per share, so they were effectively worthless. If Silver Lake did a recapitalization, we could reset the options and align interests between management and shareholders. Apple had lost most of its market share in PCs, but thanks to the iPod and iMac computers, Apple had an opportunity to reinvent itself in the consumer market. The risk/reward of investing struck me as especially favorable.

We had several conversations before Steve told me he had a better idea. He wanted me to buy up to 18 percent of Apple shares in the public market and take a board seat.

After a detailed analysis, I proposed an investment to my partners in the early fall of 2002, but they rejected it out of hand. The decision would cost Silver Lake's investors the opportunity to earn more than one hundred billion dollars in profits.

In early 2003, Bono called up with an opportunity. He wanted to buy Universal Music Group, the world's largest music label. It was a complicated transaction and took many months of analysis. A team of us did the work and presented it to my other three partners in Silver Lake in September. They agreed to do the deal with Bono, but they stipulated one condition: I would not be part of the deal team. They explained their intention for Silver Lake to go forward as a trio, rather than as a quartet. There had been signals along the way, but I had missed them. I had partnered with deal guys—people who use power when they have it to gain advantages where they can get them—and had not protected myself.

I have never believed in staying where I'm not wanted, so I quit. If I had been motivated by money, I would have hung in there, as there was no way they could force me out. I had conceived the fund, incubated it, brought in the first billion dollars of assets, and played a decisive role on the three most successful investments. But I'm not wired to fight over money. I just quit and walked out. I happened to be in New York and called Bono. He asked me to come to his apartment. When I got there, he said, "Screw them. We'll start our own fund." Elevation Partners was born.

In the long term, my departure from Silver Lake worked out for everyone. The second Silver Lake fund got off to a rocky start, as my cofounders struggled with stock picking, but they figured it out and built the firm into an institution that has delivered good investment returns to its investors.

2

Silicon Valley Before Facebook

I think technology really increased human ability.
But technology cannot produce compassion. —DALAI LAMA

The technology industry that gave birth to Facebook in 2004 bore little resemblance to the one that had existed only half a dozen years earlier. Before Facebook, startups populated by people just out of college were uncommon, and few succeeded. For the fifty years before 2000, Silicon Valley operated in a world of tight engineering constraints. Engineers never had enough processing power, memory, storage, or bandwidth to do what customers wanted, so they had to make trade-offs. Engineering and software programming in that era rewarded skill and experience. The best engineers and programmers were artists. Just as Facebook came along, however, processing power, memory, storage, and bandwidth went from being engineering limits to turbochargers of growth. The technology industry changed dramatically in less than a decade, but in ways few people recognized. What happened with Facebook and the other internet platforms could not have happened in prior generations of technology. The path the tech industry took from its founding to that change helps to explain both Facebook's success and how it could do so much damage before the world woke up.

The history of Silicon Valley can be summed in two "laws." Moore's Law, coined by a cofounder of Intel, stated that the number of transistors on an integrated circuit doubles every year. It was later revised to a more useful formulation: the performance of an integrated circuit doubles every eighteen to twenty-four months. Metcalfe's Law, named for a founder of 3Com, said that the value of any network would increase as the square of the number of nodes. Bigger networks are geometrically more valuable than small ones. Moore's Law and Metcalfe's Law reinforced each other. As the price of computers fell, the benefits of connecting them rose. It took fifty years, but we eventually connected every computer. The result was the internet we know today, a global network that connects billions of devices and made Facebook and all other internet platforms possible.

Beginning in the fifties, the technology industry went through several eras. During the Cold War, the most important customer was the government. Mainframe computers, giant machines that were housed in special air-conditioned rooms, supervised by a priesthood of technicians in white lab coats, enabled unprecedented automation of computation. The technicians communicated with mainframes via punch cards connected by the most primitive of networks. In comparison to today's technology, mainframes could not do much, but they automated large-scale data processing, replacing human calculators and bookkeepers with machines. Any customer who wanted to use a computer in that era had to accept a product designed to meet the needs of government, which invested billions to solve complex problems like moon trajectories for NASA and missile targeting for the Department of Defense. IBM was the dominant player in the mainframe era and made all the components for the machines it sold, as well as most of the software. That business model was called vertical integration. The era of government lasted about thirty years. Data networks as we think of them today did not yet exist. Even so, brilliant people imagined a world where small computers optimized for productivity would be connected on powerful networks. In the sixties, J. C. R. Licklider conceived the network that would become the internet, and he persuaded the government to finance its development. At the same time, Douglas Engelbart invented the field of human-computer interaction, which led to him to create the first computer mouse and to conceive the first graphical interface. It would take nearly two decades before Moore's Law and Metcalfe's Law could deliver enough performance to enable their vision of personal computing and an additional decade before the internet took off.

Beginning in the seventies, the focus of the tech industry began to shift toward the needs of business. The era began with a concept called time sharing, which enabled many users to share the use of a single computer, reducing the cost to everyone. Time sharing gave rise to minicomputers, which were smaller than mainframes but still staggeringly expensive by today's standards. Data networking began but was very slow and generally revolved around a single minicomputer. Punch cards gave way to terminals, keyboards attached to the primitive network, eliminating the need for a priesthood of technicians in white lab coats. Digital Equipment, Data General, Prime, and Wang led in minicomputers, which were useful for accounting and business applications but were far too complicated and costly for personal use. Although they were a big step forward relative to mainframes, even minicomputers barely scratched the surface of customer needs. Like IBM, the minicomputer vendors were vertically integrated, making most of the components for their products. Some minicomputers— Wang word processors, for example—addressed productivity applications that would be replaced by PCs. Other applications survived longer, but in the end, the minicomputer business would be subsumed by personal computer technology, if not by PCs themselves. Mainframes have survived to the present day, thanks in large part to giant, custom applications like accounting systems, which were created for the government and corporations and are cheaper to maintain on old systems than to re-create on new ones. (Massive server farms based on PC technology now attract any new application that needs mainframe-class processing; it is a much cheaper solution because you can use commodity hardware instead of proprietary mainframes.)

ARPANET, the predecessor to today's internet, began as a Department of Defense research project in 1969 under the leadership of Bob Taylor, a computer scientist who continued to influence the design of systems and networks until the late nineties. Douglas Engelbart's lab was one of the first nodes on ARPANET. The goal was to create a nation-wide network to protect the country's command and control infrastructure in the event of a nuclear attack.

The first application of computer technology to the consumer market came in 1972, when Al Alcorn created the game Pong as a training exercise for his boss at Atari, Nolan Bushnell. Bushnell's impact on Silicon Valley went far beyond the games produced by Atari. He introduced the hippie culture to tech. White shirts with pocket protectors gave way to jeans and T-shirts. Nine to five went away in favor of the crazy, but flexible hours that prevail even today.

In the late seventies, microprocessors made by Motorola, Intel, and others were relatively cheap and had enough performance to allow Altair, Apple, and others to make the first personal computers. PCs like the Apple II took advantage of the growing supply of inexpensive components, produced by a wide range of independent vendors, to deliver products that captured the imagination first of hobbyists, then of consumers and some businesses. In 1979, Dan Bricklin and Bob Frankston introduced VisiCalc, the first spreadsheet for personal computers. It is hard to overstate the significance of VisiCalc. It was an engineering marvel. A work of art. Spreadsheets on Apple IIs transformed the productivity of bankers, accountants, and financial analysts.

Unlike the vertical integration of mainframes and minicomputers, which limited product improvement to the rate of change of the slowest evolving part in the system, the horizontal integration of PCs allowed innovation at the pace of the most rapidly improving parts in the system. Because there were multiple, competing vendors for each component, systems could evolve far more rapidly than equivalent products subject to vertical integration. The downside was that PCs assembled this way lacked the tight integration of mainframes and minicomputers. This created a downstream cost in terms of training and maintenance, but that was not reflected in the purchase price and did not trouble customers. Even IBM took notice.

When IBM decided to enter the PC market, it abandoned vertical integration and partnered with a range of third-party vendors, including Microsoft for the operating system and Intel for the microprocessor. The first IBM PC shipped in 1981, signaling a fundamental change in the tech industry that only became obvious a couple of years later, when Microsoft's and Intel's other customers started to compete with IBM. Eventually, Compaq, Hewlett-Packard, Dell, and others left IBM in the dust. In the long run, though, most of the profits in the PC industry went to Microsoft and Intel, whose control of the brains and heart of the device and willingness to cooperate forced the rest of the industry into a commodity business.

ARPANET had evolved to become a backbone for regional networks of universities and the military. PCs continued the trend of smaller, cheaper computers, but it took nearly a decade after the introduction of the Apple II before technology emerged to leverage the potential of clusters of PCs. Local area networks (LANs) got their start in the late eighties as a way to share expensive laser printers. Once installed, LANs attracted developers, leading to new applications, such as electronic mail. Business productivity and engineering applications created incentives to interconnect LANs within buildings and then tie them all together over proprietary wide area networks (WANs) and then the internet. The benefits of connectivity overwhelmed the frustration of incredibly slow networks, setting the stage for steady improvement. It also created a virtuous cycle, as PC technology could be used to design and build better components, increasing the performance of new PCs that could be used to design and build even better components.

Consumers who wanted a PC in the eighties and early nineties had to buy one created to meet the needs of business. For consumers, PCs were relatively expensive and hard to use, but millions bought and learned to operate them. They put up with character-mode interfaces until Macintosh and then Windows finally delivered graphical interfaces that did not, well, totally suck. In the early nineties, consumer-centric PCs optimized for video games came to market.

The virtuous cycle of Moore's Law for computers and Metcalfe's Law for networks reached a new level in the late eighties, but the open internet did not take off right away. It required enhancements. The English researcher Tim Berners-Lee delivered the goods when he invented the World Wide Web in 1989 and the first web browser in 1991, but even those innovations were not enough to push the internet into the mainstream. That happened when a computer science student by the name of Marc Andreessen created the Mosaic browser in 1993. Within a year, startups like Yahoo and Amazon had come along, followed in 1995 by eBay, and the web that we now know had come to life.

By the mid-nineties, the wireless network evolved to a point that enabled widespread adoption of cell phones and alphanumeric pagers. The big applications were phone calls and email, then text messaging. The consumer era had begun. The business era had lasted nearly twenty years—from 1975 to 1995—but no business complained when it ended. Technology aimed at consumers was cheaper and somewhat easier to use, exactly what businesses preferred. It also rewarded a dimension that had not mattered to business: style. It took a few years for any vendor to get the formula right.

The World Wide Web in the mid-nineties was a beautiful thing. Idealism and utopian dreams pervaded the industry. The prevailing view was that the internet and World Wide Web would make the world more democratic, more fair, and more free. One of the web's best features was an architecture that inherently delivered net neutrality: every site was equal. In that first generation, everything on the web revolved around pages, every one of which had the same privileges and opportunities. Unfortunately, the pioneers of the internet made omissions that would later haunt us all. The one that mattered most was the choice not to require real identity. They never imagined that anonymity would lead to problems as the web grew.

Time would expose the naïveté of the utopian view of the internet, but at the time, most participants bought into that dream. Journalist Jenna Wortham described it this way: "The web's earliest architects and pioneers fought for their vision of freedom on the Internet at a time when it was still small forums for conversation and text-based gaming. They thought the web could be adequately governed by its users without their need to empower anyone to police it." They ignored early signs of trouble, such as toxic interchanges on message boards and in comments sections, which they interpreted as growing pains, because the potential for good appeared to be unlimited. No company had to pay the cost of creating the internet, which in theory enabled anyone to have a website. But most people needed tools for building websites, applications servers and the like. Into the breach stepped the "open source" community, a distributed network of programmers who collaborated on projects that created the infrastructure of the internet. Andreessen came out of that community. Open source had great advantages, most notably that its products delivered excellent functionality, evolved rapidly, and were free. Unfortunately, there was one serious problem with the web and open source products: the tools were not convenient or easy to use. The volunteers of the open source community had one motivation: to build the open web. Their focus was on performance and functionality, not convenience or ease of use. That worked well for the infrastructure at the heart of the internet, but not so much for consumer-facing applications.

The World Wide Web took off in 1994, driven by the Mosaic/Netscape browser and sites like Amazon, Yahoo, and eBay. Businesses embraced the web, recognizing its potential as a better way to communicate with other businesses and consumers. This change made the World Wide Web geometrically more valuable, just as Metcalfe's Law predicted. The web dominated culture in the late nineties, enabling a stock market bubble and ensuring near-universal adoption. The dot-com crash that began in early 2000 left deep scars, but the web continued to grow. In this second phase of the web, Google emerged as the most important player, organizing and displaying what appeared to be all the world's information. Apple broke the code on tech style—their products were a personal statement—and rode the consumer wave to a second life. Products like the iMac and iPod, and later the iPhone and iPad, restored Apple to its former glory and then some. At this writing, Apple is one of the most valuable companies in the world. (Fortunately, Apple is also the industry leader in protecting user privacy, but I will get to that later.)

In the early years of the new millennium, a game changing model challenged the page-centric architecture of the World Wide Web. Called Web 2.0, the new architecture revolved around people. The pioneers of Web 2.0 included people like Mark Pincus, who later founded Zynga; Reid Hoffman, the founder of LinkedIn; and Sean Parker, who had co-founded the music file sharing company Napster. After Napster, Parker launched a startup called Plaxo, which put address books in the cloud. It grew by spamming every name in every address book to generate new users, an idea that would be copied widely by social media platforms that launched thereafter. In the same period, Google had a brilliant insight: it saw a way to take control of a huge slice of the open internet. No one owned open source tools, so there was no financial incentive to make them attractive for consumers. They were designed by engineers, for engineers, which could be frustrating to non-engineers.

Google saw an opportunity to exploit the frustration of consumers and some business users. Google made a list of the most important things people did on the web, including searches, browsing, and email. In those days, most users were forced to employ a mix of open source and proprietary tools from a range of vendors. Most of the products did not work together particularly well, creating a friction Google could exploit. Beginning with Gmail in 2004, Google created or acquired compelling products in maps, photos, videos, and productivity applications. Everything was free, so there were no barriers to customer adoption. Everything worked together. Every app gathered data that Google could exploit. Customers loved the Google apps. Collectively, the Google family of apps replaced a huge portion of the open World Wide Web. It was as though Google had unilaterally put a fence around half of a public park and then started commercializing it.

The steady march of technology in the half century prior to 2000 produced so much value—and so many delightful surprises—that the industry and customers began to take positive outcomes for granted. Technology optimism was not equivalent to the law of gravity, but engineers, entrepreneurs, and investors believed that everything they did made the world a better place. Most participants bought into some form of the internet utopia. What we did not realize at the time was that the limits imposed by not having enough processing power, memory, storage, and network bandwidth had acted as a governor, limiting the damage from mistakes to a relatively small number of customers. Because the industry had done so much good in the past, we all believed that everything it would create in the future would also be good. It was not a crazy assumption, but it was a lazy one that would breed hubris.

When Zuck launched Facebook in early 2004, the tech industry had begun to emerge from the downturn caused by the dot-com meltdown. Web 2.0 was in its early stages, with no clear winners. For Silicon Valley, it was a time of transformation, with major change taking place in four arenas: startups, philosophy, economics, and culture. Collectively, these changes triggered unprecedented growth and wealth creation. Once the gravy train started, no one wanted to get off. When fortunes can be made overnight, few people pause to ask questions or consider side effects.

The first big Silicon Valley change related to the economics of startups. Hurdles that had long plagued new companies evaporated. Engineers could build world-class products quickly, thanks to the trove of complementary software components, like the Apache server and the Mozilla browser, from the open source community. With open source stacks as a foundation, engineers could focus all their effort on the valuable functionality of their app, rather than building infrastructure from the ground up. This saved time and money. In parallel, a new concept emerged—the cloud—and the industry embraced the notion of centralization of shared resources. The cloud is like Uber for data— customers don't need to own their own data center or storage if a service provides it seamlessly from the cloud. Today's leader in cloud services, Amazon Web Services (AWS), leveraged Amazon.com's retail business to create a massive cloud infrastructure that it offered on a turnkey basis to startups and corporate customers. By enabling companies to outsource their hardware and network infrastructure, paying a monthly fee instead of the purchase price of an entire system, services like AWS lowered the cost of creating new businesses and shortened the time to market. Startups could mix and match free open source applications to create their software infrastructure. Updates were made once, in the cloud, and then downloaded by users, eliminating what had previously been a very costly and time-consuming process of upgrading individual PCs and servers. This freed startups to focus on their real value added, the application that sat on top of the stack.

Netflix, Box, Dropbox, Slack, and many other businesses were built on this model.

Thus began the "lean startup" model. Without the huge expense and operational burden of creating a full tech infrastructure, new companies did not have to aim for perfection when they launched a new product, which had been Silicon Valley's primary model to that point. For a fraction of the cost, they could create a minimum viable product (MVP), launch it, and see what happened. The lean startup model could work anywhere, but it worked best with cloud software, which could be updated as often as necessary. The first major industry created with the new model was social media, the Web 2.0 startups that were building networks of people rather than pages. Every day after launch, founders would study the data and tweak the product in response to customer feedback. In the lean startup philosophy, the product is never finished. It can always be improved. No matter how rapidly a startup grew, AWS could handle the load, as it demonstrated in supporting the phenomenal growth of Netflix. What in earlier generations would have required an army of experienced engineers could now be accomplished by relatively inexperienced engineers with an email to AWS. Infrastructure that used to require a huge capital investment could now be leased on a monthly basis. If the product did not take off, the cost of failure was negligible, particularly in comparison to the years before 2000. If the product found a market, the founders had alternatives. They could raise venture capital on favorable terms, hire a bigger team, improve the product, and spend to acquire more users. Or they could do what the founders of Instagram and WhatsApp would eventually do: sell out for billions with only a handful of employees.

Facebook's motto—"Move fast and break things"—embodies the lean startup philosophy. Forget strategy. Pull together a few friends, make a product you like, and try it in the market. Make mistakes, fix them, repeat. For venture investors, the lean startup model was a godsend. It allowed venture capitalists to identify losers and kill them before they burned through much cash. Winners were so valuable that a fund needed only one to provide a great return.

When hardware and networks act as limiters, software must be elegant. Engineers sacrifice frills to maximize performance. The no-frills design of Google's search bar made a huge difference in the early days, providing a competitive advantage relative to Excite, Altavista, and Yahoo. A decade earlier, Microsoft's early versions of Windows failed in part because hardware in that era could not handle the processing demands imposed by the design. By 2004, every PC had processing power to spare. Wired networks could handle video. Facebook's design outperformed MySpace in almost every dimension, providing a relative advantage, but the company did not face the fundamental challenges that had prevailed even a decade earlier. Engineers had enough processing power, storage, and network bandwidth to change the world, at least on PCs. Programming still rewarded genius and creativity, but an entrepreneur like Zuck did not need a team of experienced engineers with systems expertise to execute a business plan. For a founder in his early twenties, this was a lucky break. Zuck could build a team of people his own age and mold them. Unlike Google, Facebook was reluctant to hire people with experience. Inexperience went from being a barrier to being an advantage, as it kept labor costs low and made it possible for a young man in his twenties to be an effective CEO. The people in Zuck's inner circle bought into his vision without reservation, and they conveyed that vision to the rank-and-file engineers. On its own terms, Facebook's human resources strategy worked exceptionally well. The company exceeded its goals year after year, creating massive wealth for its shareholders, but especially for Zuck. The success of Facebook's strategy had a profound impact on the human resources culture of Silicon Valley startups.

In the early days of Silicon Valley, software engineers generally came from the computer science and electrical engineering programs at

MIT, Caltech, and Carnegie Mellon. By the late seventies, Berkeley and Stanford had joined the top tier. They were followed in the mid-nineties by the University of Illinois at Urbana-Champaign, the alma mater of Marc Andreessen, and other universities with strong computer science programs. After 2000, programmers were coming from just about every university in America, including Harvard.

When faced with a surplus for the first time, engineers had new and exciting options. The wave of startups launched after 2003 could have applied surplus processing, memory, storage, and bandwidth to improve users' well-being and happiness, for example. A few people tried, which is what led to the creation of the Siri personal assistant, among other things. The most successful entrepreneurs took a different path. They recognized that the penetration of broadband might enable them to build global consumer technology brands very quickly, so they opted for maximum scale. To grow as fast as possible, they did everything they could to eliminate friction like purchase prices, criticism, and regulation. Products were free, criticism and privacy norms ignored. Faced with the choice between asking permission or begging forgiveness, entrepreneurs embraced the latter. For some startups, challenging authority was central to their culture. To maximize both engagement and revenues, Web 2.0 startups focused their technology on the weakest elements of human psychology. They set out to create habits, evolved habits into addictions, and laid the groundwork for giant fortunes.

The second important change was philosophical. American business philosophy was becoming more and more proudly libertarian, nowhere more so than in Silicon Valley. The United States had beaten the Depression and won World War II through collective action. As a country, we subordinated the individual to the collective good, and it worked really well. When the Second World War ended, the US economy prospered by rebuilding the rest of the world. Among the many peacetime benefits was the emergence of a prosperous middle class. Tax rates were high, but few people complained. Collective action enabled the country to build the best public education system in the world, as well as the interstate highway system, and to send men to the moon. The average American enjoyed an exceptionally high standard of living.

Then came the 1973 oil crisis, when the Organization of Petroleum Exporting Countries initiated a boycott of countries that supported Israel in the Yom Kippur War. The oil embargo exposed a flaw in the US economy: it was built on cheap oil. The country had lived beyond its means for most of the sixties, borrowing aggressively to pay for the war in Vietnam and the Great Society social programs, which made it vulnerable. When rising oil prices triggered inflation and economic stagnation, the country transitioned into a new philosophical regime.

The winner was libertarianism, which prioritized the individual over the collective good. It might be framed as "you are responsible only for yourself." As the opposite of collectivism, libertarianism is a philosophy that can trace its roots to the frontier years of the American West. In the modern context, it is closely tied to the belief that markets are always the best way to allocate resources. Under libertarianism, no one needs to feel guilty about ambition or greed. Disruption can be a strategy, not just a consequence. You can imagine how attractive a philosophy that absolves practitioners of responsibility for the impact of their actions on others would be to entrepreneurs and investors in Silicon Valley. They embraced it. You could be a hacker, a rebel against authority, and people would reward you for it. Unstated was the leverage the philosophy conferred on those who started with advantages. The well-born and lucky could attribute their success to hard work and talent, while blaming the less advantaged for not working hard enough or being untalented. Many libertarian entrepreneurs brag about the "meritocracy" inside their companies. Meritocracy sounds like a great thing, but in practice there are serious issues with Silicon Valley's version of it. If contributions to corporate success define merit when a company is small and has a homogeneous employee base, then meritocracy will encourage the hiring of people with similar backgrounds and experience. If the company is not careful, this will lead to a homogeneous workforce as the company grows. For internet platforms, this means an employee base consisting overwhelmingly of white and Asian males in their twenties and thirties. This can have an impact on product design. For example, Google's facial-recognition software had problems recognizing people of color, possibly reflecting a lack of diversity in the development team. Homogeneity narrows the range of acceptable ideas and, in the case of Facebook, may have contributed to a work environment that emphasizes conformity. The extraordinary lack of diversity in Silicon Valley may reflect the pervasive embrace of libertarian philosophy. Zuck's early investor and mentor Peter Thiel is an outspoken advocate for libertarian values.

The third big change was economic, and it was a natural extension of libertarian philosophy. Neoliberalism stipulated that markets should replace government as the rule setter for economic activity. President Ronald Reagan framed neoliberalism with his assertion that "government is not the solution to our problem; it is the problem." Beginning in 1981, the Reagan administration began removing regulations on business. He restored confidence, which unleashed a big increase in investment and economic activity. By 1982, Wall Street bought into the idea, and stocks began to rise. Reagan called it Morning in America. The problems—stagnant wages, income inequality, and a decline in startup activity outside of tech—did not emerge until the late nineties.

Deregulation generally favored incumbents at the expense of startups. New company formation, which had peaked in 1977, has been in decline ever since. The exception was Silicon Valley, where large companies struggled to keep up with rapidly evolving technologies, creating opportunities for startups. The startup economy in the early eighties was tiny but vibrant. It grew with the PC industry, exploded in the nineties, and peaked in 2000 at $120 billion, before declining by 87 percent over two years. The lean startup model collapsed the cost of startups, such that the number of new companies rebounded very quickly. According to the National Venture Capital Association, venture funding recovered to seventy-nine billion dollars in 2015 on 10,463 deals, more than twice the number funded in 2008. The market power of Facebook, Google, Amazon, and Apple has altered the behavior of investors and entrepreneurs, forcing startups to sell out early to one of the giants or crowd into smaller and less attractive opportunities.

Under Reagan, the country also revised its view of corporate power. The Founding Fathers associated monopoly with monarchy and took steps to ensure that economic power would be widely distributed. There were ebbs and flows as the country adjusted to the industrial revolution, mechanization, technology, world wars, and globalization, but until 1981, the prevailing view was that there should be limits to the concentration of economic power and wealth. The Reagan Revolution embraced the notion that the concentration of economic power was not a problem so long as it did not lead to higher prices for consumers. Again, Silicon Valley profited from laissez-faire economics.

Technology markets are not monopolies by nature. That said, every generation has had dominant players: IBM in mainframes, Digital Equipment in minicomputers, Microsoft and Intel in PCs, Cisco in data networking, Oracle in enterprise software, and Google on the internet. The argument against monopolies in technology is that major innovations almost always come from new players. If you stifle the rise of new companies, innovation may suffer.

Before the internet, the dominant tech companies sold foundational technologies for the architecture of their period. With the exception of Digital Equipment, all of the tech market leaders of the past still exist today, though none could prevent their markets from maturing, peaking, and losing ground to subsequent generations. In two cases, IBM and Microsoft, the business practices that led to success eventually caught the eye of antitrust regulators, resulting in regulatory actions that restored competitive balance. Without the IBM antitrust case, there likely would have been no Microsoft. Without the Microsoft case, it is hard to imagine Google succeeding as it did. Beginning with Google, the most successful technology companies sat on top of stacks created by others, which allowed them to move faster than any market leaders before them. Google, Facebook, and others also broke the mold by adopting advertising business models, which meant their products were free to use, eliminating another form of friction and protecting them from antitrust regulation. They rode the wave of wired broadband adoption and then 4G mobile to achieve global scale in what seemed like the blink of an eye. Their products enjoyed network effects, which occur when the value of a product increases as you add users to the network. Network effects were supposed to benefit users. In the cases of Facebook and Google, that was true for a time, but eventually the value increase shifted decisively to the benefit of owners of the network, creating insurmountable barriers to entry. Facebook and Google, as well as Amazon, quickly amassed economic power on a scale not seen since the days of Standard Oil one hundred years earlier. In an essay on Medium, the venture capitalist James Currier pointed out that the key to success in the internet platform business is network effects and Facebook enjoyed more of them than any other company in history. He said, "To date, we've actually identified that Facebook has built no less than six of the thirteen known network effects to create defensibility and value, like a castle with six concentric layers of walls. Facebook's walls grow higher all the time, and on top of them Facebook has fortified itself with *all three* of the other known defensibilities in the internet age: brand, scale, and embedding."

By 2004, the United States was more than a generation into an era dominated by a hands-off, laissez-faire approach to regulation, a time period long enough that hardly anyone in Silicon Valley knew there had once been a different way of doing things. This is one reason why few people in tech today are calling for regulation of Facebook, Google, and Amazon, antitrust or otherwise.

One other factor made the environment of 2004 different from earlier times in Silicon Valley: angel investors. Venture capitalists had served as the primary gatekeepers of the startup economy since the late seventies, but they spent a few years retrenching after the dot-com bubble burst. Into the void stepped angel investors—individuals, mostly former entrepreneurs and executives—who guided startups during their earliest stages. Angel investors were perfectly matched to the lean startup model, gaining leverage from relatively small investments. One angel, Ron Conway, built a huge brand, but the team that had started PayPal proved to have much greater impact. Peter Thiel, Elon Musk, Reid Hoffman, Max Levchin, Jeremy Stoppleman, and their colleagues were collectively known as the PayPal Mafia, and their impact transformed Silicon Valley. Not only did they launch Tesla, Space-X, LinkedIn, and Yelp, they provided early funding to Facebook and many other successful players. More important than the money, though, were the vision, value system, and connections of the PayPal Mafia, which came to dominate the social media generation. Validation by the PayPal Mafia was decisive for many startups during the early days of social media. Their management techniques enabled startups to grow at rates never before experienced in Silicon Valley. The value system of the PayPal Mafia helped their investments create massive wealth, but may have contributed to the blindness of internet platforms to harms that resulted from their success. In short, we can trace both the good and the bad of social media to the influence of the PayPal Mafia.

THANKS TO LUCKY TIMING, Facebook benefitted not only from lower barriers for startups and changes in philosophy and economics but also from a new social environment. Silicon Valley had prospered in the suburbs south of San Francisco, mostly between Palo Alto and San Jose. Engineering nerds did not have a problem with life in the sleepy suburbs because many had families with children, and the ones who did not have kids did not expect to have the option of living in the city. Beginning with the dot-com bubble of the late nineties, however, the startup culture began to attract kids fresh out of school, who were not so happy with suburban life as their predecessors. In a world where experience had declining economic value, the new generation favored San Francisco as a place to live. The transition was bumpy, as most of the San Francisco–based dot-coms went up in flames in 2000, but after the start of the new millennium, the tech population in San Francisco grew steadily. While Facebook originally based itself in Palo Alto—the heart of Silicon Valley, not far from Google, Hewlett-Packard, and Apple—a meaningful percentage of its employees chose to live in the big city. Had Facebook come along during the era of scarcity, when experienced engineers ruled the Valley, it would have had a profoundly different culture. Faced with the engineering constraints of earlier eras, however, the Facebook platform would not have worked well enough to succeed. Facebook came along at the perfect time.

San Francisco is hip, with diverse neighborhoods, decent public transportation, access to recreation, and lots of nightlife. It attracted a different kind of person than Sunnyvale or Mountain View, including two related types previously unseen in Silicon Valley: hipsters and bros. Hipsters had burst onto the public consciousness as if from a base in Brooklyn, New York, heavy on guys with beards, plaid shirts, and earrings. They seemed to be descendants of San Francisco's bohemian past, a modern take on the Beats. The bros were different, though perhaps more in terms of style than substance. Ambitious, aggressive, and exceptionally self-confident, they embodied libertarian values. Symptoms included a lack of empathy or concern for consequences to others. The hipster and bro cultures were decidedly male. There were women in tech, too, more than in past generations of Silicon Valley, but the culture continued to be dominated by men who failed to appreciate the obvious benefits of treating women as peers. Too many in Silicon Valley missed the lesson that treating others as equals is what good people do. For them, I make a simple economic case: women are 51 percent of the US population; they account for 85 percent of consumer purchases; they control 60 percent of all personal wealth. They know what they want better than men do, yet in Silicon Valley, which invests billions in consumer-facing startups, men hold most of the leadership positions. Women who succeed often do so by beating the boys at their own game, something that Silicon Valley women do with ever greater frequency. *Bloomberg* journalist Emily Chang described this culture brilliantly in her book, *Brotopia*.

With the biggest influx of young people since the Summer of Love, the tech migration after 2000 had a visible impact on the city, precipitating a backlash that began quietly but grew steadily. The new kids boosted the economy with tea shops and co-working spaces that sprung up like mushrooms after a summer rain in the forest. But they seemed not to appreciate that their lifestyle might disturb the quiet equilibrium that had preceded their arrival. With a range of new services catering to their needs, delivered by startups of their peers, the hipsters and bros eventually provoked a reaction. Tangible manifestations of their presence, like the luxury buses that took them to jobs at Google, Facebook, Apple, and other companies down in Silicon Valley, drew protests from peeved locals. An explosion of Uber and Lyft vehicles jammed the city's streets, dramatically increasing commute times. Insensitive blog posts, inappropriate business behavior, and higher housing costs ensured that locals would neither forgive nor forget.

ZUCK ENJOYED THE KIND OF privileged childhood one would expect for a white male whose parents were medical professionals living in a beautiful suburb. As a student at Harvard, he launched Facebook. Thanks to great focus and enthusiasm, Zuck would almost certainly have found success in Silicon Valley in any era, but he was particularly suited to his times. Plus, as previously noted, he had an advantage not available to earlier generations of entrepreneurs: he could build a team of people his age—many of whom had never before had a full-time job—and mold them. This allowed Facebook to accomplish things that had never been done before.

For Zuck and the senior management of Facebook, the goal of connecting the world was self-evidently admirable. The philosophy of "move fast and break things" allowed for lots of mistakes, and Facebook embraced the process, made adjustments, and continued forward. The company maintained a laser focus on Zuck's priorities, never considering the possibility that there might be flaws in this approach, even when the evidence of such flaws became overwhelming. From all appearances, Zuck and his executive team did not anticipate that people would use Facebook differently than Zuck had envisioned, that putting more than two billion people on the same network would lead to tribalism, that Facebook Groups would amplify that tribalism, that bad actors would take advantage to harm innocent people. They failed to imagine unintended consequences from an advertising business based on behavior modification. They ignored critics. They missed the opportunity to take responsibility when the reputational cost would have been low. When called to task, they protected their business model and prerogatives, making only small changes to their business practices. This trajectory is worth understanding in greater depth.

3

Move Fast and Break Things

Try not to become a man of success, but rather
try to become a man of value. —ALBERT EINSTEIN

During Mark Zuckerberg's sophomore year at Harvard, he created a program called Facemash that allowed users to compare photos of two students and choose which was "hotter." The photos were taken from the online directories of nine Harvard dormitories. According to an article in *Fast Company* magazine, the application had twenty-two thousand photo views in the first four hours and spread rapidly on campus before being shut down within a week by the authorities. Harvard threatened to expel Zuckerberg for security, copyright, and privacy violations. The charges were later dropped. The incident caught the attention of three Harvard seniors, Cameron Winklevoss, Tyler Winklevoss, and Divya Narendra, who invited Zuck to consult on their social network project, HarvardConnection.com.

In an interview with the campus newspaper, Zuck complained that the university would be slow to implement a universal student directory and that he could do it much faster. He started in January 2004 and launched TheFacebook.com on February 4. Six days later, the trio of seniors accused Zuck of pretending to help on their project and then stealing their ideas for TheFacebook. (The Winklevoss twins and Narendra ultimately filed suit and settled in 2008 for 1.2 million shares of Facebook stock.) Within a month, more than half of the Harvard student body had registered on Zuck's site. Three of Zuck's friends joined the team, and a month later they launched TheFacebook at Columbia, Stanford, and Yale. It spread rapidly to other college campuses. By June, the company relocated from Cambridge, Massachusetts, to Palo Alto, California, brought in Napster cofounder Sean Parker as president, and took its first venture capital from Peter Thiel.

TheFacebook delivered exactly what its name described: each page provided a photo with personal details and contact information. There was no News Feed and no frills, but the color scheme and fonts would be recognizable to any present-day user. While many features were missing, the thing that stands out is the effectiveness of the first user interface. There were no mistakes that would have to be undone.

The following year, Zuck and team paid two hundred thousand dollars to buy the "facebook.com" domain and changed the company's name. Accel Partners, one of the leading Silicon Valley venture funds, invested $12.7 million, and the company expanded access to high school students and employees of some technology firms. The functionality of the original Facebook was the same as TheFacebook, but the user interface evolved. Some of the changes were subtle, such as the multitone blue color scheme, but others, such as the display of thumbnail photos of friends, remain central to the current look. Again, Facebook made improvements that would endure. Sometimes users complained about new features and products—this generally occurred when Zuck and his team pushed users too hard to disclose and share more information—but Facebook recovered quickly each time. The company never looked back.

Facebook was not the first social network. SixDegrees.com started in 1997 and Makeoutclub in 1999, but neither really got off the ground. Friendster, which started in 2002, was the first to reach one million users. Friendster was the model for Facebook. It got off to a fantastic start, attracted investors and users, but then fell victim to performance problems that crippled the business. Friendster got slower and slower, until users gave up and left the platform. Started in 2003, MySpace figured out how to scale better than Friendster, but it, too, eventually had issues. Allowing users to customize pages made the system slow, but in the end, it was the ability of users to remain anonymous that probably did the most damage to MySpace. Anonymity encouraged the posting of pornography, the elimination of which drained MySpace's resources, and enabled adults to pose as children, which led to massive problems.

The genius of Zuck and his original team was in reconceptualizing the problem. They recognized that success depended on building a network that could scale without friction. Sean Parker described the solution this way in Adam Fisher's *Valley of Genius*: "The 'social graph' is a math concept from graph theory, but it was a way of trying to explain to people who were kind of academic and mathematically inclined that what we were building was not a product so much as it was a network composed of nodes with a lot of information flowing between those nodes. That's graph theory. Therefore we're building a social graph. It was never meant to be talked about publicly." Perhaps not, but it was brilliant. The notion that a small team in their early twenties with little or no work experience figured it out on the first try is remarkable. The founders also had the great insight that real identity would simplify the social graph, reducing each user to a single address. These two ideas would not only help Facebook overcome the performance problems that sank Friendster and MySpace, they would lay the foundation for a company with more than two billion users.

When I first met Zuck in 2006, I was very familiar with Friendster and MySpace and had a clear sense that Facebook's design, its insistence on real identity, and user control of privacy would enable the company to succeed where others had failed. Later on, Facebook would relax its policies on identity and privacy to enable faster growth. Facebook's terms of service still require real identity, but enforcement is lax, consistent with the company's commitment to minimize friction, and happens only when other users complain. By the end of the decade, user privacy would become a pawn to be traded to accelerate growth.

In 2006, it was not obvious how big the social networking market would be, but I was already convinced that Facebook had an approach that might both define the category and make it economically successful. Facebook was a hit with college students, but I thought the bigger opportunity would be with adults, whose busy schedules were tailor-made for the platform. To me, that suggested a market opportunity of at least one hundred million users or more in English-speaking countries. In those days, one hundred million users would have justified a valuation of at least ten billion dollars, or ten times the number Yahoo had offered. It never occurred to me then that Facebook would fly past two billion monthly users, though I do remember the first time Zuck told me his target was a billion users. It happened some time in 2009, when Facebook was racing from two hundred to three hundred million users. I thought it was a mistake to maximize user count. The top 20 percent of users would deliver most of the value. I worried that the pursuit of one billion users would force Zuck to do business in places or on terms that should make him uncomfortable. As it turned out, there were no visible compromises when Facebook passed a billion monthly users in September 2012. The compromises were very well hidden.

The company had plenty of capital when I first met Zuck, so there was no immediate opportunity for me to invest, but as I've said, the notion of helping the twenty-two-year-old founder of a game-changing startup deal with an existential crisis really appealed to me. As a long-time technology investor, I received many requests for free help, and I loved doing it. Good advice can be the first step in a lasting relationship and had ultimately led to many of my best investments. The strategy required patience—and a willingness to help lots of companies that might not work out—but it made my work life fresh and fun.

My first impression of Zuck was that he was a classic Silicon Valley nerd. In my book, being a nerd is a good thing, especially for a technology entrepreneur. Nerds are my people. I didn't know much about Zuck as a person and knew nothing about the episode that nearly led to his expulsion from Harvard until much later. What I saw before me was a particularly intense twenty-two-year-old who took all the time he needed to think before he acted. As painful as that five minutes of silence was for me, it signaled caution, which I took as a positive. The long silence also signaled weak social skills, but that would not have been unusual in a technology founder. But in that first meeting, I was able to help Zuck resolve a serious problem. Not only did he leave my office with the answer he needed, he had a framework for justifying it to the people in his life who wanted their share of one billion dollars. At the time, Zuck was very appreciative. A few days later, he invited me to his office, which was in the heart of Palo Alto, just down the street from the Stanford University campus. The interior walls were covered with graffiti. Professional graffiti. In Zuck's conference room, we talked about the importance of having a cohesive management team where everyone shared the same goals. Those conversations continued several times a month for three years. Thanks to the Yahoo offer, Zuck understood that he could no longer count on everyone on his team. Some executives had pushed hard to sell the company. Zuck asked for my perspective on team building, which I was able to provide in the course of our conversations. A year later, he upgraded several positions, most notably his chief operating officer and his chief financial officer.

Toward the end of 2006, Zuck learned that a magazine for Harvard alumni was planning a story about the Winklevoss brothers and again turned to me for help. I introduced him to a crisis-management public relations firm and helped him minimize the fallout from the story.

I trust my instincts about people. My instincts are far from perfect, but they have been good enough to enable a long career. Intensity of the kind I saw in Zuck is a huge positive in an entrepreneur. Another critical issue for me is a person's value system. In my interactions with him, Zuck was consistently mature and responsible. He seemed remarkably grown-up for his age. He was idealistic, convinced that Facebook could bring people together. He was comfortable working with women, which is not common among Silicon Valley entrepreneurs. My meetings with Zuck almost always occurred in his office, generally just the two of us, so I had an incomplete picture of the man, but he was always straight with me. I liked Zuck. I liked his team. I was a fan of Facebook.

This is a roundabout way of saying that my relationship with Zuck was all business. I was one of the people he would call on when confronted with new or challenging issues. Mentoring is fun for me, and Zuck could not have been a better mentee. We talked about stuff that was important to Zuck, where I had useful experience. More often than not, he acted on my counsel.

Zuck had other mentors, several of whom played a much larger role than I did. He spoke to me about Peter Thiel, who was an early investor and board member. I don't know how often Zuck spoke with Thiel, but I know he took Peter's advice very seriously. Philosophically, Thiel and I are polar opposites, and I respected Zuck for being able to work with both of us. *Washington Post* CEO Don Graham had started advising Zuck at least a year before me. As one of the best-connected people in our nation's capital, Don would have been a tremendous asset to Zuck as Facebook grew to global scale. Marc Andreessen, the Netscape founder turned venture capitalist, played a very important role in Zuck's orbit, as he was a hard-core technologist who had once been a very young entrepreneur. Presumably, Zuck also leaned on Jim Breyer, the partner from Accel who made the first institutional investment in Facebook, but Zuck did not talk about Breyer the way he did about Thiel.

In researching this book for key moments in the history of Facebook, one that stands out occurred months before I got involved. In the fall of 2005, Facebook gave users the ability to upload photographs. They did it with a new wrinkle—tagging the people in the photo—that helped to define Facebook's approach to engagement. Tagging proved to be a technology with persuasive power, as users felt obligated to react or reciprocate when informed they had been tagged. A few months after my first meeting with Zuck, Facebook made two huge changes: it launched News Feed, and it opened itself up to anyone over the age of thirteen with a valid email address. News Feed is the heart of the Facebook user experience, and it is hard today to imagine that the site did well for a couple of years without it. Then, in January 2007, Facebook introduced a mobile web product to leverage the widespread adoption of smartphones. The desktop interface also made a big leap.

In the summer of 2007, Zuck called to offer me an opportunity to invest. He actually offered me a choice: invest or join the board. Given my profession and our relationship, the choice was easy. I did not need to be on the board to advise Zuck. The investment itself was complicated. One of Facebook's early employees needed to sell a piece of his stake, but under the company's equity-incentive plan there was no easy way to do this. We worked with Facebook to create a structure that balanced both our needs and those of the seller. When the deal was done, there was no way to sell our shares until after an initial public offering. Bono, Marc, and I were committed for the long haul.

Later that year, Microsoft bought 1.6 percent of Facebook for $240 million, a transaction that valued the company at $15 billion. The transaction was tied to a deal where Microsoft would sell advertising for Facebook. Microsoft paid a huge premium to the price we paid, reflecting its status as a software giant with no ability to compete in social. Facebook understood that it had leverage over Microsoft and priced the shares accordingly. As investors, we knew the Microsoft valuation did not reflect the actual worth of Facebook. It was a "strategic investment" designed to give Microsoft a leg up over Google and other giants.

Soon thereafter, Facebook launched Beacon, a system that gathered data about user activity on external websites to improve Facebook ad targeting and to enable users to share news about their purchases. When a Facebook user interacted with a Beacon partner website, the data would be sent to Facebook and reflected in the user's News Feed. Beacon was designed to make Facebook advertising much more valuable, and Facebook hoped that users would be happy to share their interests and purchase activities with friends. Unfortunately, Facebook did not give users any warning and did not give them any ability to control Beacon. Their activities on the web would appear in their Facebook feed even when the user was not on Facebook. Imagine having "Just looked at sex toys on Amazon.com" show up in your feed. Users thought Beacon was creepy. Most users did not know what Facebook was doing with Beacon. When they found out, they were not happy. Zuck's cavalier attitude toward user privacy, evident from the first day of Facemash back at Harvard, had blown up in his face. MoveOn organized a protest campaign, arguing that Facebook should not publish user activity off the site without explicit permission. Users filed class action lawsuits. Beacon was withdrawn less than a year after launch.

In the fall of 2007, Zuck told me he wanted to hire someone to build Facebook's monetization. I asked if he was willing to bring in a strong number two, someone who could be a chief operating officer or president. He said yes. I did not say anything, but a name sprang to mind immediately: Sheryl Sandberg. Sheryl had been chief of staff to Secretary of the Treasury Larry Summers during Bill Clinton's second term. In that job, she had partnered with Bono on the singer's successful campaign to spur the world's leading economies to forgive billions in debt owed by countries in the developing world. Together, Bono and Sheryl helped many emerging countries to reenergize their economies, which turned out to be a good deal for everyone involved. Sheryl introduced Bono to me, which eventually led the two of us to collaborate on Elevation Partners. Sheryl came to Silicon Valley in early 2001 and hung out in my office for a few weeks. We talked to Sheryl about joining Integral, but my partner John Powell had a better idea. John and I were both convinced that Sheryl would be hugely successful in Silicon Valley, but John pointed out that there were much bigger opportunities than Integral. He thought the right place for Sheryl was Google and shared that view with John Doerr, who was a member of Google's board of directors. Sheryl took a job at Google to help build AdWords, the product that links ads to search results.

AdWords is arguably the most successful advertising product in history, and Sheryl was one of the people who made that happen. Based on what I knew about Sheryl, her success came as no surprise. One day in 2007, Sheryl came by to tell me she had been offered a leadership position at *The Washington Post*. She asked me what I thought. I suggested that she consider Facebook instead. Thanks to Watergate and the Pentagon Papers, the *Post* was iconic, but being a newspaper, it did not have a workable plan to avoid business model damage from the internet. Facebook seemed like a much better match for Sheryl than the *Post*, and she seemed like the best possible partner for Zuck and Facebook. Sheryl told me she had once met Zuck at a party, but did not know him and worried that they might not be a good fit. I encouraged Sheryl to get to know Zuck and see where things went. After my first conversation with Sheryl, I called Zuck and told him I thought Sheryl would be the best person to build Facebook's advertising business. Zuck worried that advertising on Facebook would not look like Google's AdWords—which was true—but I countered that building AdWords might be the best preparation for creating a scalable advertising model on Facebook. It took several separate conversations with Zuck and Sheryl to get them to meet, but once they got together, they immediately found common ground. Sheryl joined the company in March 2008. Looking at a March 2008 *Wall Street Journal* article on Sheryl's hire and Zuck's other efforts to stabilize the company by accepting help from more experienced peers, I'm reminded that Facebook's current status as a multibillion-dollar company seemed far from inevitable in those days. The article highlighted the company's image problems and mentioned Zuck complaining to me about the difficulties of being a CEO. Still, growth accelerated.

The underlying technology of the disastrous Beacon project resurfaced in late 2008 as Facebook Connect, a product that allowed users to sign into third-party sites with their Facebook credentials. News of hacks and identity theft had created pressure for stronger passwords, which users struggled to manage. The value of Connect was that it enabled people to memorize a single, strong Facebook password for access to thousands of sites. Users loved Connect for its convenience, but it is not obvious that they understood that it enabled Facebook to track them in many places around the web. With the benefit of hindsight, we can see the costs that accompanied the convenience of Connect. I tried Connect on a few news sites, but soon abandoned it when I realized what it meant for privacy.

The data that Facebook collected through Connect led to huge improvements in targeting and would ultimately magnify catastrophes like the Russian interference in the 2016 election. Other users must have noticed that Facebook knew surprising things about them, but may have told themselves the convenience of Connect justified the loss of privacy. With Connect, Facebook addressed a real need. Maintaining secure credentials is inconvenient, but the world would have been better off had users adopted a solution that did not exploit their private data. Convenience, it turns out, was the sweetener that led users to swallow a lot of poison.

Facebook's user count reached one hundred million in the third quarter of 2008. This was astonishing for a company that was only four and half years old, but Facebook was just getting started. Only seven months later, the user count hit two hundred million, aided by the launch of the Like button. The Like button soon defined the Facebook experience. "Getting Likes" became a social phenomenon. It gave users an incentive to spend more time on the site and joined photo tagging as a trigger for addiction to Facebook. To make its advertising valuable, Facebook needs to gain and hold user attention, which it does with behavior modification techniques that promote addiction, according to a growing body of evidence. Behavior modification and addiction would play a giant role in the Facebook story, but were not visible during my time as a mentor to Zuck, and I would not appreciate their significance until 2017.

It turns out everyone wants to be liked, and the Like button provided a yardstick of social validation and social reciprocity—packaged as a variable reward—that transformed social networking. It seemed that every Facebook user wanted to know how many Likes they received for each post, and that tempted many users to return to the platform several times a day. Facebook amplified the signal with notifications, teasing users constantly. The Like button helped boost the user count to 305 million by the end of September 2009. Like buttons spread like wildfire to sites across the web, and along with Connect enabled Facebook to track its users wherever they browsed.

The acquisition of FriendFeed in August 2009 gave Facebook an application for aggregating feeds from a wide range of apps and blogs. It also provided technology and a team that would protect Facebook's flank from the new kid on the block, Twitter. Over the following year, Facebook acquisitions would enable photo sharing and the importing of contacts. Such acquisitions made Facebook more valuable to users, but that was nothing compared to the value they created for Facebook's advertising. On every metric, Facebook prospered. Revenue grew rapidly. Facebook's secret sauce was its ability to imitate and improve upon the ideas of others, and then scale them. The company demonstrated an exceptional aptitude for managing hypergrowth, a skill that is as rare as it is valuable. In September 2009, the company announced that it had turned cash flow positive. This is not the same as turning profitable, but it was actually a more important milestone. It meant that Facebook generated enough revenue to cover all its cash expenses. It would not need more venture capital to survive. The company was only five and a half years old.

With Sheryl on board as chief operating officer in charge of delivering revenues, Facebook quickly developed its infrastructure to enable rapid growth. This simplified Zuck's life so he could focus on strategic issues. Facebook had transitioned from startup to serious business. This coming-of-age had implications for me, too. Effectively, Zuck had graduated. With Sheryl as his partner, I did not think Zuck would need mentoring from me any longer. My domain expertise in mobile made me valuable as a strategy advisor, but even that would be a temporary gig. Like most successful entrepreneurs and executives, Zuck is brilliant (and ruthless) about upgrading his closest advisors as he goes along. In the earliest days of Facebook, Sean Parker played an essential role as president, but his skills stopped matching the company's needs, so Zuck moved on from him. He also dropped the chief operating officer who followed Parker and replaced him with Sheryl. The process is Darwinian in every sense. It is natural and necessary. I have encountered it so many times that I can usually anticipate the right moment to step back. I never give it a moment's thought.

Knowing that we had accomplished everything we could have hoped for at the time I began mentoring him, I sent Zuck a message saying that my job was done. He was appreciative and said we would always be friends. At this point, I stopped being an insider, but I remained a true believer in Facebook. While failures like Beacon had foreshadowed problems to come, all I could see was the potential of Facebook as a force for good. The Arab Spring was still a year away, but the analyst in me could see how Facebook might be used by grassroots campaigns. What I did not grasp was that Zuck's ambition had no limit. I did not appreciate that his focus on code as the solution to every problem would blind him to the human cost of Facebook's outsized success. And I never imagined that Zuck would craft a culture in which criticism and disagreement apparently had no place.

The following year, 2010, was big for Facebook in surprising ways. By July, Facebook had five hundred million users, half of whom visited the site every day. Average daily usage was thirty-four minutes. Users who joined Facebook to stay in touch with family soon found new functions to enjoy. They spent more time on the site, shared more posts, and saw more ads.

October saw the release of *The Social Network*, a feature film about the early days of Facebook. The film was a critical and commercial success, winning three Academy Awards and four Golden Globes. The plot focused on Zuck's relationship with the Winklevoss twins and the lawsuit that resulted from it. The portrayal of Zuck was unflattering. Zuck complained that the film did not accurately tell the story, but hardly anyone besides him seemed to care. I chose not to watch the film, preferring the Zuck I knew to a version crafted in Hollywood.

Just before the end of 2010, Facebook improved its user interface again, edging closer to the look and feel we know today. The company finished 2010 with 608 million monthly users. The rate of user growth remained exceptionally high, and minutes of use per user per day continued to rise. Early in 2011, Facebook received an investment of five hundred million dollars for 1 percent of the company, pushing the valuation up to fifty billion dollars. Unlike the Microsoft deal, this transaction reflected a financial investor's assessment of Facebook's value. At this point, even Microsoft was making money on its investment. Facebook was not only the most exciting company since Google, it showed every indication that it would become one of the greatest tech companies of all time. New investors were clamoring to buy shares. By June 2011, DoubleClick announced that Facebook was the most visited site on the web, with more than one trillion visits. Nielsen disagreed, saying

Facebook still trailed Google, but it appeared to be only a matter of time before the two companies would agree that Facebook was #1.

In March 2011, I saw a presentation that introduced the first seed of doubt into my rosy view of Facebook. The occasion was the annual TED Conference in Long Beach, the global launch pad for TED Talks. The eighteen-minute Talks are thematically organized over four days, providing brain candy to millions far beyond the conference. That year, the highlight for me was a nine-minute talk by Eli Pariser, the board president of MoveOn.org. Eli had an insight that his Facebook and Google feeds had stopped being neutral. Even though his Facebook friend list included a balance of liberals and conservatives, his tendency to click more often on liberal links had led the algorithms to prioritize such content, eventually crowding out conservative content entirely. He worked with friends to demonstrate that the change was universal on both Facebook and Google. The platforms were pretending to be neutral, but they were filtering content in ways that were invisible to users. Having argued that the open web offered an improvement on the biases of traditional content editors, the platforms were surreptitiously implementing algorithmic filters that lacked the value system of human editors. Algorithms would not act in a socially responsible way on their own. Users would think they were seeing a balance of content when in fact they were trapped in what Eli called a "filter bubble" created and enforced by algorithms. He hypothesized that giving algorithms gatekeeping power without also requiring civic responsibility would lead to unexpected, negative consequences. Other publishers were jumping on board the personalization bandwagon. There might be no way for users to escape from filter bubbles.

Eli's conclusion? If platforms are going to be gatekeepers, they need to program a sense of civic responsibility into their algorithms. They need to be transparent about the rules that determine what gets through the filter. And they need to give users control of their bubble.

I was gobsmacked. It was one of the most insightful talks I had ever heard. Its import was obvious. When Eli finished, I jumped out of my seat and made a beeline to the stage door so that I could introduce myself. If you view the talk today, you will immediately appreciate its importance. At the time I did not see a way for me to act on Eli's insight at Facebook. I no longer had regular contact with Zuck, much less inside information. I was not up to speed on the engineering priorities that had created filter bubbles or about plans for monetizing them. But Eli's talk percolated in my mind. There was no good way to spin filter bubbles. All I could do was hope that Zuck and Sheryl would have the sense not to use them in ways that would harm users. (You can listen to Eli Pariser's "Beware Online 'Filter Bubbles'" talk for yourself on TED.com.)

Meanwhile, Facebook marched on. Google introduced its own social network, Google+, in June 2011, with considerable fanfare. By the time Google+ came to market, Google had become a gatekeeper between content vendors and users, forcing content vendors who wanted to reach their own audience to accept Google's business terms. Facebook took a different path to a similar place. Where most of Google's products delivered a single function that gained power from being bundled, Facebook had created an integrated platform, what is known in the industry as a walled garden, that delivered many forms of value. Some of the functions on the platform had so much value that Facebook spun them off as stand-alone products. One example: Messenger.

Thanks to its near monopoly of search and the AdWords advertising platform that monetized it, Google knew more about purchase intentions than any other company on earth. A user looking to buy a hammer would begin with a search on Google, getting a set of results along with three AdWords ads from vendors looking to sell hammers. The search took milliseconds. The user bought a hammer, the advertiser sold one, and Google got paid for the ad. Everyone got what they wanted. But Google was not satisfied. It did not know the consumer's identity. Google realized that its data set of purchase intent would have greater value if it could be tied to customer identity. I call this McNamee's 7th Law: data sets become geometrically more valuable when you combine them. That is where Gmail changed the game. Users got value in the form of a good email system, but Google received something far more valuable. By tying purchase intent to identity, Google laid the foundation for new business opportunities. It then created Google Maps, enabling it to tie location to purchase intent and identity. The integrated data set rivaled Amazon's, but without warehouses and inventory it generated much greater profits for Google. Best of all, combined data sets often reveal insights and business opportunities that could not have been imagined previously. The new products were free to use, but each one contributed data that transformed the value of Google's advertising products. Facebook did something analogous with each function it added to the platform. Photo tagging expanded the social graph. News Feed enriched it further. The Like button delivered data on emotional triggers. Connect tracked users as they went around the web. The value is not really in the photos and links posted by users. The real value resides in metadata—data about data—which is what we call the data that describes where the user was when he or she posted, what they were doing, with whom they were doing it, alternatives they considered, and more. Broadcast media like television, radio, and newspapers lack the real-time interactivity necessary to create valuable metadata. Thanks to metadata, Facebook and Google create a picture of the user that can be monetized more effectively than traditional media. When collected on the scale of Google and Facebook, metadata has unimaginable value. When people say, "In advertising businesses, users are not the customer; they are the product," this is what they are talking about. But in the process, Facebook in particular changed the nature of advertising. Traditional advertising seeks to persuade, but in a one-size-fits-most kind of way. The metadata that Facebook and others collected enabled them to find unexpected patterns, such as "four men who collect baseball cards, like novels by Charles Dickens, and check Facebook after midnight bought a certain model of Toyota," creating an opportunity to package male night owls who collect baseball cards and like Dickens for car ads. Facebook allows advertisers to identify each user's biases and appeal to them individually. Insights gathered this way changed the nature of ad targeting. More important, though, all that data goes into Facebook's (or Google's) artificial intelligence and can be used by advertisers to exploit the emotions of users in ways that increase the likelihood that they purchase a specific model of car or vote in a certain way. As the technology futurist Jaron Lanier has noted, advertising on social media platforms has evolved into a form of manipulation.

Google+ was Google's fourth foray into social networking. Why did Google try so many times? Why did it keep failing? By 2011, it must have been obvious to Google that Facebook had the key to a new and especially valuable online advertising business. Unlike traditional media or even search, social networking provided signals about each user's emotional state and triggers. Relative to the monochrome of search, social network advertising offered Technicolor, the equivalent of Oz vs. Kansas in *The Wizard of Oz*. If you are trying to sell a commodity product like a hammer, search advertising is fine, but for branded products like perfume or cars or clothing, social networking's data on emotions has huge incremental value. Google wanted a piece of that action. Google+ might have added a new dimension to Google's advertising business, but Facebook had a prohibitive lead when Google+ came to market, and the product's flaws prevented it from gaining much traction with people outside of Google. All it offered was interesting features, and Facebook imitated the good parts quickly.

Facebook took no chances with Google+. The company went to battle stations and devoted every resource to stopping Google on the beach of social networking. The company cranked up its development efforts, dramatically increasing the size limits for posts, partnering with Skype, introducing the Messenger texting product, and adding a slew of new tools for creating applications on the platform. As 2012 began, Facebook was poised for a breakout year. The company had a new advertising product—Open Graph—that leveraged its Social Graph, the tool to capture everything it knew from both inside Facebook and around the web. Initially, Facebook gave advertisers access only to data captured inside the platform. Facebook also enabled advertisements in the News Feed for the first time. News Feed ads really leveraged Facebook's user experience. Ads blended in with posts from friends, which meant more people saw them, but there was also a downside: it was very hard to get an ad to stand out the way it would on radio or TV or in print.

The big news early in 2012 came when Facebook filed for an initial public offering (IPO) and then acquired Instagram for one billion dollars. The Facebook IPO, which took place on May 17, raised sixteen billion dollars, making it the third largest in US history at the time. The total valuation of $104 billion was the highest ever for a newly public company. Facebook had revenues of nearly four billion dollars and net income of one billion dollars in the year prior to the IPO and found itself in the Fortune 500 list of companies from day one.

As impressive as all those numbers are, the IPO itself was something of a train wreck. Trading glitches occurred during the first day, preventing some trades from going through, and the stock struggled to stay above the IPO price. The deal set a record for trading volume on the first day after an IPO: 460 million shares.

The months leading up to the IPO saw weakness in Facebook's advertising sales that triggered reductions in the company's revenue forecast. When a company is preparing for an IPO, forecast reductions can be disastrous, as public investors have no incentive to buy into uncertainty. In Facebook's case, investors' extreme enthusiasm for the company—based primarily on user growth and Facebook's increasing impact on society—meant the IPO could survive the reduction in forecast, but Zuck's dream of a record-setting offering might be at risk. As described by former Facebook advertising targeting manager Antonio García Martínez in his book *Chaos Monkeys*, "The narratives the company had woven about the new magic of social-media marketing were in deep reruns with advertisers, many of whom were beginning to openly question the fortunes they had spent on Facebook thus far, often with little to show for it." For all its success with users, Facebook had not yet created an advertising product that provided the targeting necessary to provide appropriate results for advertisers. Martínez went on to say, "A colossal yearlong bet the company had made on a product called Open Graph, and its accompanying monetization spin-off, Sponsored Stories, had been an absolute failure in the market." Advertisers had paid a lot of money to Facebook, believing the company's promises about ad results, but did not get the value they felt they deserved. For Facebook, this was a moment of truth. By pushing the IPO valuation to record levels, Facebook set itself up for a rocky start as a public company.

The newly public stock sold off almost immediately and went into free fall after Yahoo Finance reported that the investment banks that had underwritten the IPO had reduced their earnings forecasts just before the offering. In the heat of the deal, had those forecast changes been effectively communicated to buyers of the stock? The situation was sufficiently disturbing that regulatory authorities initiated a review. Lawsuits followed, alleging a range of violations with respect to the trading glitches and the actions of one underwriter. A subsequent set of lawsuits named the underwriters, Zuck and Facebook's board, and Nasdaq. *The Wall Street Journal* characterized the IPO as a "fiasco."

For Facebook's business, though, the IPO was an undisputed blessing. The company received a staggering amount of free publicity before the deal, essentially all of it good. That turbocharged user growth, news of which enabled Facebook to survive the IPO issues with relatively little damage. Investors trusted that a company with such impressive user growth would eventually figure out monetization. Once again, Facebook pushed the envelope, stumbled, and got away with it. Then they did something really aggressive.

The data from inside Facebook alone did not deliver enough value for advertisers. Thanks to Connect and the ubiquitous Like and Share buttons, Facebook had gathered staggering amounts of data about user behavior from around the web. The company had chosen not to use the off-site data for commercial purposes, a self-imposed rule that it decided to discard when the business slowed down. No one knew yet how valuable the external data would be, but they decided to find out. As Martínez describes it, Zuck and Sheryl began cautiously, fearful of alienating users.

Thanks to the IPO, Facebook enjoyed a tsunami of user growth. Within a few months, user growth restored investor confidence. It also overwhelmed the complaints from advertisers, who had to go where their customers were, even if the ad vehicles on Facebook were disappointing. The pressure to integrate user data from activities away from Facebook into the ad products lessened a bit, but the fundamental issues with targeting and the value of ads remained. As a result, the decision to integrate user data from outside Facebook would not be reversed.

In early October 2012, the company announced it had surpassed one billion monthly users, with 600 million mobile users, 219 billion photo uploads, and 140 billion friend connections. Despite the mess of the IPO—and not being privy to the issues with ads—I took great pride in Facebook's success. The stock turned out to be a game changer for Elevation. Even though my partners had turned down our first opportunity to invest, Elevation subsequently made a large investment at a relatively low price, ensuring on its own that the fund would be a winner.

Only eight and a half years from Zuck's dorm room, Facebook had become a powerful economic engine. Thanks to the philosophy of "move fast and break things," no one at Facebook was satisfied with a record-setting IPO. They began hacking away at the problem of monetizing users. There were several challenges. As Martínez wrote in *Chaos Monkeys*, the advertising team around the time of the IPO was, for the most part, young people who had no previous work experience in advertising or even media. They learned everything by trial and error. For every innovation, there were many mistakes, some of which would have been obvious to a more experienced team. The team may have been young, but they were smart, highly motivated, and persistent. Their leadership, with Sheryl Sandberg at the top, created a successful sales culture. They took a long view and learned from every mistake. They focused on metrics.

In the early days, Facebook did its best to create effective advertising products and tools from profile data, users' friend relationships, and user actions on the site. My band, Moonalice, was an early advertiser, with a budget of less than ten thousand dollars a year. Our first ads, a few years before the IPO, were tiny rectangles on the side of the page, with a few words of text and maybe a link. The goal was to introduce Moonalice to new fans. We promoted a song called "It's 4:20 Somewhere" this way. We ran an ad for several years—typically spending ten or twenty dollars a day—and people downloaded the song 4.6 million times, a number cited by the Rock & Roll Hall of Fame as a record from any band's own website. A little Facebook ad running every day for three years made it possible. But when the only option was a tiny rectangle, Facebook ads were ineffective for many advertisers and products. The same format that worked so well for downloading a song was worthless for promoting a conference. I have no idea why it worked for one thing and not the other. It did not matter much, because in those days, Facebook allowed plenty of free distribution, which they called "organic reach." Back then, for a fan page likes ours, Facebook would let a really compelling post reach about 15 percent of our fans for free. The value of organic reach on Facebook compelled us and millions of others to shift the focus of our communications from a website to Facebook. We trained our fans to interact with us on Facebook and to use our website as a content archive. Many others did the same, helping to cement Facebook's position as the social hub of the web. Embracing Facebook worked really well for Moonalice, and our page eventually gathered more than 420,000 followers.

Not surprisingly, there was a catch. Every year or so, Facebook would adjust the algorithm to reduce organic reach. The company made its money from advertising, and having convinced millions of organizations to set up shop on the platform, Facebook held all the cards. The biggest beneficiaries of organic reach had no choice but to buy ads to maintain their overall reach. They had invested too much time and had established too much brand equity on Facebook to abandon the platform. Organic reach declined in fits and starts until it finally bottomed at about 1 percent or less. Fortunately, Facebook would periodically introduce a new product—the Facebook Live video service, for example—and give those new products greater organic reach to persuade people like us to use them. We signed up for Facebook Live on the first day it was available and streamed a concert that same day. The reach was fantastic. Facebook Live and Moonalice were made for each other. I streamed one set of the very first concert by Dead & Company, a spin-off from the Grateful Dead, and so many people watched it that the band saw a surge in ticket sales for other dates on the tour. As a result, one of the band's managers invited me to stream their next show from the stage.

At the time of the IPO, targeting options for Facebook ads were limited to demographic information from activity on the site, things like age, sex, and location, as well as interests and relationships. The introduction of Open Graph and News Feed ads in 2012 set the stage for much better targeting, which improved rapidly when Facebook integrated "off-site" data into the suite available to advertisers. If Moonalice wanted to promote a concert, we would target demographic data—say, people over twenty-one in the city where our gig was—and then filter that audience with interests, like "concerts," "Beatles," and "hippie." We would spend perhaps one hundred dollars to promote a show on Facebook. We would get a few thousand "impressions," which in theory meant users who saw the ad. The nature of News Feed is such that users race past a lot of posts. To capture attention, we switched from promoting Events—which is what Facebook called our concerts—to creating posts that included a rich graphic element, such as a poster. In doing so, we ran afoul of the 20 percent rule. As it was explained to me by a senior Facebook executive, Zuck had decided that too much text made ads boring, so he set an arbitrary limit of 20 percent text. Our posters are works of art, but many of them violated the 20 percent rule because rock posters sometimes integrate lots of text into the art. Facebook would reject those ads, so I learned to superimpose a little bit of text onto attention-grabbing photos to create compelling images that would comply with Facebook's rules.

Moonalice is not a sophisticated advertiser, but we were not alone in that regard. Facebook enabled millions of organizations with small budgets to reach audiences for a fraction of the cost of print, radio, or TV advertisements. But Facebook recognized that the really big money would come from attracting the advertisers who had historically spent giant budgets on traditional media. Such advertisers had completely different expectations. They expected to reach large, targeted audiences at a reasonable cost with complete transparency, which is to say they wanted proof that their messages reached their intended audiences. At the time of the IPO, Facebook could not meet those expectations consistently. In 2013, Facebook began experimenting with data from user activity outside Facebook. They created tools for advertisers to exploit that data. The tools enabled Facebook advertisers the ability to target audiences whose emotions were being triggered in predictable ways.

Facebook's culture matched its advertising challenge perfectly. A company that prided itself on its software hacking roots perfected a new model to monetize its success. Growth hacking applies the intensely focused, iterative model of software hacking to the problem of increasing user count, time on site, and revenue. It works only when a company has a successful product and a form of monetization that can benefit from tinkering, but for the right kind of company, growth hacking can be transformational. Obsessive focus on metrics is a central feature of growth hacking, so it really matters that you pick the correct metrics.

From late 2012 to 2017, Facebook perfected growth hacking. The company experimented constantly with algorithms, new data types, and small changes in design, measuring everything. Every action a user took gave Facebook a better understanding of that user—and of that user's friends—enabling the company to make tiny improvements in the "user experience" every day, which is to say they got better at manipulating the attention of users. The goal of growth hacking is to generate more revenue and profits, and at Facebook those metrics blocked out all other considerations. In the world of growth hacking, users are a metric, not people. It is unlikely that civic responsibility ever came up in Facebook's internal conversations about growth hacking. Once the company started applying user data from outside the platform, there was no turning back. The data from outside Facebook transformed targeting inside Facebook. Additional data improved it more, giving Facebook an incentive to gather data anywhere it could be gathered. The algorithms looked for and found unexpected correlations in the data that could be monetized effectively. Before long, Facebook's surveillance capabilities rivaled those of an intelligence agency.

To deliver better targeting, Facebook introduced new tools for advertisers. Relative to the 2016 election, the two most important may have been Custom Audiences and Lookalike Audiences. As described in Facebook's Advertiser Help Center, a Custom Audience is "a type of audience you can create made up of your existing customers. You can target ads to the audience you've created on Facebook, Instagram, and Audience Network," the last of which is "a network of publisher-owned apps and sites where you can show your ads" outside the Facebook platform. Introduced in 2013, Custom Audiences had two very important forms of value to advertisers: first, advertisers could build an ad campaign around known customers, and second, a Custom Audience could be used to create a Lookalike Audience, which finds other Facebook users who share characteristics with the Custom Audience. Lookalike scales without limit, so advertisers have the option to find every user on Facebook with a given set of characteristics. You could start with as few as one hundred people, but the larger the Custom Audience, the better the Lookalike will be. Facebook recommends using a Custom Audience between one thousand and fifty thousand.

Thanks to growth hacking, Facebook made continuous improvements in its advertising tools, as well as growing its audience, increasing time on site, and gathering astonishing amounts of data. Progress against these metrics translated into explosive revenue growth. From one billion users at year-end 2012, Facebook grew to 1.2 billion in 2013, 1.4 billion in 2014, 1.6 billion in 2015, nearly 1.9 billion in 2016, 2.1 billion in 2017, 2.3 billion in 2018, and 2.4 billion in late 2019. From just more than $5 billion in sales in the IPO year of 2012, Facebook grew to $7.8 billion in 2013, $12.5 billion in 2014, $17.9 billion in 2015, $27.6 billion in 2016, $40.7 billion in 2017, and $55.8 billion in 2018. There were issues along the way. Advertisers complained about the lack of transparency relative to advertising, and Facebook has been sued for inflating some metrics, including ad views and video views. But Facebook had become a juggernaut. The customers that advertisers needed to reach were on Facebook. This gave Facebook enormous leverage. When advertisers complained, Facebook could get away with apologies and marginal fixes. With its customary laser focus on a handful of metrics, Facebook did not devote any energy to questioning its decisions. If there was any soul searching about the morality of intense surveillance and the manipulation of user attention, or about protecting users against unintended consequences, I have been able to find no evidence of it. If Zuck and the Facebook team noticed that usage of Facebook differed materially from their ideal, they showed no concern. If anyone noticed the increasingly extreme behavior inside some Facebook Groups, no one took action. The Russians exploited this to sow dissention among Americans and Western Europeans, beginning in 2014. When *The Guardian* newspaper in the UK broke the story in December 2015 that Cambridge Analytica had misappropriated profiles from at least fifty million Facebook users, it precipitated an intense but brief scandal. Facebook apologized and made Cambridge Analytica sign a piece of paper, certifying that it had destroyed the data set, but then quickly returned to business as usual. While always careful to protect itself from legal liability, Facebook seemed oblivious to signs of trouble. The benefits of growth—in terms of revenue, profits, and influence— were obvious, the problems easy to ignore. At Facebook, everyone remained focused on their metrics.

As 2016 began, Facebook was on a huge roll. Aside from a few PR headaches, the company had not skipped a beat since its IPO. Almost everything that mattered had changed since my days as a mentor to Zuck, and I knew only what had been publicly disclosed or what I had seen with my own eyes. Like its users, Facebook showed the public only the good news, which is why I was so surprised by what I saw during 2016. Bad actors using Facebook's tools to harm innocent people did not compute for me, but I saw the evidence and could not let it go. It was only when I reengaged with Zuck and Sheryl just before the election that I began to appreciate that my view of Facebook was inaccurate. It took me longer than I would have liked to understand the problem. More than four years of relentless success at Facebook had bred overconfidence. The company was in a filter bubble of its own. Every day, there were more users, spending more time on the site, generating more revenue and earnings, which pushed the stock to new highs. The Midas Effect may have set in, causing Zuck and his team to believe that everything they did was right, always for the best, and uncontestably good for humanity. Humility went out the window. Facebook subordinated everything to growth. Eventually that would create problems it could not resolve with an apology and a promise to do better.

4

The Children of Fogg

*It's not because anyone is evil or has bad
intentions. It's because the game is getting
attention at all costs.* —TRISTAN HARRIS

On April 9, 2017, onetime Google design ethicist Tristan Harris appeared on *60 Minutes* with Anderson Cooper to discuss the techniques that internet platforms like Facebook, Twitter, YouTube, Instagram, and Snapchat use to prey on the emotions of their users. He talked about the battle for attention among media, how smartphones transformed that battle, and how internet platforms profit from that transformation at the expense of their users. The platforms prey on weaknesses in human psychology, using ideas from propaganda, public relations, and slot machines to create habits, then addiction. Tristan called it "brain hacking."

By the time I saw Tristan's interview, I had spent three months unsuccessfully trying to persuade Facebook that its business model and algorithms were a threat to users and also to its own brand. I realized I couldn't do it alone—I needed someone who could help me understand what I had observed over the course of 2016. Tristan's vision explained so much of what I had seen. His focus was on public health, but I saw immediately the implications for elections and economics.

I got Tristan's contact information and called him the next day. He told me that he had been trying to get engineers at technology companies like Google to understand brain hacking for more than three years. We decided to join forces. Our goal was to make the world aware of the dark side of social media. Our focus would be on Tristan's public health framework, but we would look for opportunities to address political issues, such as elections, and economic issues like innovation and entrepreneurship. Our mission might be quixotic, but we were determined to give it a try.

Tristan was born in 1984, the year of the Macintosh. He grew up as the only child of a single mother in Santa Rosa, California, an hour or so north of the Golden Gate Bridge and San Francisco. When Tristan got his first computer at age five, he fell in love. As a child, Tristan showed particular interest in magic, going to a special camp where his unusual skills led to mentoring by several professional magicians. As performed by magicians, magic tricks exploit the evolutionary foundations of human attention. Just as all humans smile more or less the same way, we also respond to certain visual stimuli in predictable ways. Magicians know a lot about how attention works, and they structure their tricks to take advantage. That's how a magician's coin appears to fly from one hand to the other and then disappears. Or how a magician can make a coin disappear and then reappear from a child's ear. When a magician tells you to "pick a card, any card," they do so after a series of steps designed to cause you to pick a very specific card. All these tricks work on nearly every human because they play on our most basic wiring. We cannot help but be astonished because our attention has been manipulated in an unexpected manner. Language, culture, and even education level do not matter to a magician. The vast majority of humans react the same way.

For young Tristan, magic gave way to computers in the transition from elementary to middle school. Computers enabled Tristan to build stuff, which seemed like magic. He embraced programming languages the way some boys embrace baseball stats, making games and applications of increasing sophistication. It was the late nineties, and Apple was just emerging from a slump that had lasted more than a decade. Tristan fell in love with his Mac and with Apple, dreaming of working there one day. It didn't take long, thanks to the admissions department at Stanford.

Stanford University is the academic hub of Silicon Valley. Located less than two hours south of Tristan's home in Santa Rosa, Stanford has given birth to many of the most successful technology companies in history, including Google. When Tristan arrived at Stanford in the fall of 2002, he focused on computer science. Less than a month into his freshman year, he followed through on his dream and applied for a summer internship at Apple, which he got, working mostly on design projects. Some of the code and the user interfaces he created over the course of three summer jobs remain in Apple products today.

After graduation, Tristan enrolled in the graduate computer science master's program at Stanford. In his first term, he took a class in persuasive technology with Professor B. J. Fogg, whose textbook, *Persuasive Technology*, is the standard in the field. Professors at other universities teach the subject, but being at Stanford gave Fogg outsized influence in Silicon Valley. His insight was that computing devices allow programmers to combine psychology and persuasion concepts from the early twentieth century, like propaganda, with techniques from slot machines, like variable rewards, and tie them to the human social need for approval and validation in ways that few users can resist. Like a magician doing a card trick, the computer designer can create the illusion of user control when it is the system that guides every action. Fogg's textbook lays out a formula for persuasion that clever programmers can exploit more effectively on each new generation of technology to hijack users' minds. Prior to smartphones like the iPhone and Android, the danger was limited. After the transition to smartphones, users did not stand a chance. Fogg did not help. As described in his textbook, Fogg taught ethics by having students "work in small teams to develop a conceptual design for an ethically questionable persuasive technology—the more unethical the better." He thought this was the best way to get students to think about the consequences of their work.

Disclosure that the techniques he taught may have contributed to undermining democracy and public health have led to criticism of Professor Fogg himself. After reading Fogg's textbook and a Medium post he wrote, I developed a sense that he is a technology optimist who embraced Silicon Valley's value system, never imagining that his insights might lead to material harm. I eventually had an opportunity to speak to Fogg. He is a thoughtful and friendly man who feels he is being unfairly blamed for the consequences of persuasive technology on internet platforms. He told me that he made several attempts to call attention to the dangers of persuasive technology, but that Silicon Valley paid no attention.

In companies like Facebook and Google, Fogg's disciples often work in what is called the Growth group, the growth hackers charged with increasing the number of users, time on site, and engagement with ads. They have been very successful. When we humans interact with internet platforms, we think we are looking at cat videos and posts from friends in a simple news feed. What few people know is that behind the news feed is a large and advanced artificial intelligence. When we check a news feed, we are playing multidimensional chess against massive artificial intelligences that have nearly perfect information about us. The goal of the AI is to figure out which content will keep each of us highly engaged and monetizable. Success leads the AI to show us more content like whatever engaged us in the past. For the 1.47 billion users who check Facebook every day, reinforcement of beliefs, every day for a year or two, will have an effect. Not on every user in every case, but on enough users in enough situations to be both effective for advertising and harmful to democracy.

The artificial intelligences of companies like Facebook (and Google) now include behavioral prediction engines that anticipate our thoughts and emotions, based on patterns found in the reservoir of data they have accumulated about users. Years of Likes, posts, shares, comments, and Groups have taught Facebook's AI how to monopolize our attention. Thanks to all this data, Facebook can offer advertisers exceptionally high-quality targeting. The challenge has been to create ad products that extract maximum value from that targeting.

The battle for attention requires constant innovation. As the industry learned with banner ads in the early days of the internet, users adapt to predictable ad layouts, skipping over them without registering any of the content. When it comes to online ads, there's a tradeoff. On the one hand, it is a lot easier to make sure the right person is seeing your ad. On the other hand, it is a lot *harder* to make sure that person is *paying attention* to the ad. For the tech platforms, the solution to the latter problem is to maximize the time users spend on the platform. If they devote only a small percentage of their attention to the ads they see, then the key is to monopolize as much of their attention as possible. So Facebook (and other platforms) add new content formats and products in the hope of stimulating more engagement. In the beginning, text was enough. Then photos took over. Then mobile. Video is the new frontier. In addition to new formats, Facebook also introduces new products, such as Messenger and a dating service. To maximize profits, internet platforms, including Facebook, hide the ball on the effectiveness of ads.

Platforms provide less-than-industry-standard visibility to advertisers, preventing traditional audit practices. The effectiveness of advertising has always been notoriously difficult to assess—hence the aphorism "I know half my ad spending is wasted; I just don't know which half"—and platform ads work well enough that advertisers generally spend more every year. Search ads on Google offer the clearest payback; brand ads on other platforms are much harder to measure. What matters, though, is that advertisers need to put their message in front of prospective customers, no matter where they may be. As users gravitate from traditional media to the internet, the ad dollars follow them. Until they come up with an ad format that is truly compelling, platforms will do whatever they can to maximize daily users and time on site. So long as the user is on the site, the platform will get paid for ads.

INTERNET PLATFORMS HAVE EMBRACED B. J. Fogg's approach to persuasive technology, applying it in every way imaginable on their sites. Autoplay and endless feeds eliminate cues to stop. Unpredictable, variable rewards stimulate behavioral addiction. Tagging, Like buttons, and notifications trigger social validation loops. As users, we do not stand a chance. Humans have evolved a common set of responses to certain stimuli—"flight or fight" would be an example—that can be exploited by technology. When confronted with visual stimuli, such as vivid colors—red is a trigger color—or a vibration against the skin near our pocket that signals a possible enticing reward, the body responds in predictable ways: a faster heartbeat and the release of a neurotransmitter, dopamine. In human biology, a faster heartbeat and the release of dopamine are meant to be momentary responses that increase the odds of survival in a life-or-death situation. Too much of that kind of stimulus is a bad thing for any human, but the effects are particularly dangerous in children and adolescents. The first wave of consequences includes lower sleep quality, an increase in stress, anxiety, depression, an inability to concentrate, irritability, and insomnia. That is just the beginning. Many of us develop nomophobia, which is the fear of being separated from one's phone. We are conditioned to check our phones constantly, craving ever more stimulation from our platforms of choice. Many of us develop problems relating to and interacting with other people. Kids get hooked on games, texting, Instagram, and Snapchat that change the nature of human experience. Cyberbullying becomes easy over texting and social media because when technology mediates human relationships, the social cues and feedback loops that would normally cause a bully to experience shunning or disgust by their peers are not present. Adults get locked into filter bubbles, which Wikipedia defines as "a state of intellectual isolation that can result from personalized searches when a website algorithm selectively guesses what information a user would like to see based on information about the user, such as location, past click-behavior and search history." Filter bubbles promote engagement, which makes them central to the business models of Facebook and Google. But filter bubbles are not unique to internet platforms. They can also be found on any journalistic medium that reinforces the preexisting beliefs of its audience, while suppressing any stories that might contradict them. Partisan TV channels like Fox News and MSNBC maintain powerful filter bubbles, but they cannot match the impact of Facebook and Google because television is a one-way, broadcast medium. It does not allow for personalization, interactivity, sharing, or groups.

In the context of Facebook, filter bubbles have several elements. In the endless pursuit of engagement, Facebook's AI and algorithms feed each of us a steady diet of content similar to what has engaged us most in the past. Usually that is content we "like." Every click, share, and comment helps Facebook refine its algorithms just a little bit. With 2.4 billion people clicking, sharing, and commenting every month—1.58 billion every day—Facebook's AI knows more about users than they can imagine. All that data in one place would be a target for bad actors, even if it were well-protected. But Facebook's business model is to give the opportunity to exploit that data to just about anyone who is willing to pay for the privilege.

Tristan makes the case that platforms compete in a race to the bottom of the brain stem—where the AIs present content that appeals to the low-level emotions of the lizard brain, things like immediate rewards, outrage, and fear. Short videos perform better than longer ones. Animated GIFs work better than static photos. Sensational headlines work better than calm descriptions of events. As Tristan says, the space of true things is fixed, while the space of falsehoods can expand freely in any direction—false outcompetes true. From an evolutionary perspective, that is a huge advantage. People say they prefer puppy photos and facts—and that may be true for many—but inflammatory posts work better at reaching huge audiences within Facebook and other platforms.

Getting a user outraged, anxious, or afraid is a powerful way to increase engagement. Anxious and fearful users check the site more frequently. Outraged users share more content to let other people know what they should also be outraged about. Best of all from Facebook's perspective, outraged or fearful users in an emotionally hijacked state become more reactive to further emotionally charged content. It is easy to imagine how inflammatory content would accelerate the heart rate and trigger dopamine hits. Facebook knows so much about each user that they can often tune News Feed to promote emotional responses. They cannot do this all the time to every user, but they do it far more than users realize. And they do it subtly, in very small increments. On a platform like Facebook, where most users check the site every day, small daily nudges over long periods of time can eventually produce big changes. In 2014, Facebook published a study called "Experimental Evidence of Massive-Scale Emotional Contagion Through Social Networks," where they manipulated the balance of positive and negative messages in the News Feeds of nearly seven hundred thousand users to measure the influence of social networks on mood. In its internal report, Facebook claimed the experiment provided evidence that emotions can spread over its platform. Without getting prior informed consent or providing any warning, Facebook made people sad just to see if it could be done. Confronted with a tsunami of criticism, Sheryl Sandberg said this: "This was part of ongoing research companies do to test different products, and that was what it was; it was poorly communicated. And for that communication we apologize. We never meant to upset you." She did not apologize for running a giant psychological experiment on users. She claimed that experiments like this are normal "for companies." And she concluded by apologizing only for Facebook's poor communication. If Sheryl's comments are any indication, running experiments on users without prior consent is a standard practice at Facebook.

It turns out that connecting 2.4 billion people on a single network does not naturally produce happiness for all. It puts pressure on users, first to present a desirable image, then to command attention in the form of Likes or shares from others. In such an environment, the loudest voices dominate, which can be intimidating. As a result, we follow the human instinct to organize ourselves into clusters or tribes. This starts with people who share our beliefs, most often family, friends, and Facebook Groups to which we belong. Facebook's News Feed enables every user to surround him- or herself with like-minded people. While Facebook notionally allows us to extend our friend network to include a highly diverse community, in practice, many users stop following people with whom they disagree. When someone provokes us, it feels good to cut them off, so lots of people do that. The result is that friends lists become more homogeneous over time, an effect that Facebook amplifies with its approach to curating News Feed. When content is coming from like-minded family, friends, or Groups, we tend to relax our vigilance, which is one of the reasons why disinformation spreads so effectively on Facebook.

Giving users what they want sounds like a great idea, but it has at least one unfortunate by-product: filter bubbles. There is a high correlation between the presence of filter bubbles and polarization. To be clear,

I am not suggesting that filter bubbles create polarization, but I believe they have a negative impact on public discourse and politics because filter bubbles isolate the people stuck in them. Filter bubbles exist outside Facebook and Google, but gains in attention for Facebook and Google are increasing the influence of their filter bubbles relative to others.

Everyone on Facebook has friends and family, but many are also members of Groups. Facebook allows Groups on just about anything, including hobbies, entertainment, teams, communities, churches, and celebrities. There are many Groups devoted to politics, across the full spectrum. Facebook loves Groups because they enable easy targeting by advertisers. Bad actors like them for the same reason. Research by Cass Sunstein, who was the administrator of the White House Office of Information and Regulatory Affairs for the first Obama administration, indicates that when like-minded people discuss issues, their views tend to get more extreme over time.

Groups of politically engaged users who share a common set of beliefs reinforce each other, provoking shared outrage at perceived enemies, which, as I previously noted, makes them vulnerable to manipulation. Jonathon Morgan of Data for Democracy has observed that as few as 1 to 2 percent of a group can steer the conversation if they are well-coordinated. That means a human troll with a small army of digital bots—software robots—can control a large, emotionally engaged Group, which is what the Russians did when they persuaded Groups on opposite sides of the same issue—like pro-Muslim groups and anti-Muslim groups—to simultaneously host Facebook events in the same place at the same time, hoping for a confrontation.

Facebook wants us to believe that it is merely a platform on which others act and that it is not responsible for what those third parties do. Both assertions warrant debate. In reality, Facebook created and operates a complex system built around a value system that increasingly conflicts with the values of the users it is supposed to serve. Where Facebook asserts that users control their experience by picking the friends and sources that populate their News Feed, in reality an artificial intelligence, algorithms, and menus created by Facebook engineers control every aspect of that experience. With nearly as many monthly users as there are notional Christians in the world, and nearly as many daily users as there are notional Muslims, Facebook cannot pretend its business model and design choices do not have a profound effect. Facebook's notion that a platform with more than two billion users can and should police itself also seems both naïve and self-serving, especially given the now plentiful evidence to the contrary. Even if it were "just a platform," Facebook has a responsibility for protecting users from harm. Deflection of responsibility has serious consequences.

THE COMPETITION FOR ATTENTION across the media and technology spectrum rewards the worst social behavior. Extreme views attract more attention, so platforms recommend them. News Feeds with filter bubbles do better at holding attention than News Feeds that don't have them. If the worst thing that happened with filter bubbles was that they reinforced preexisting beliefs, they would be no worse than many other things in society. Unfortunately, people in a filter bubble become increasingly tribal, isolated, and extreme. They seek out people and ideas that make them comfortable.

Social media has enabled personal views that had previously been kept in check by social pressure—white nationalism is an example—to find an outlet. Before the platforms arrived, extreme views were often moderated because it was hard for adherents to find one another. Expressing extreme views in the real world can lead to social stigma, which also keeps them in check. By enabling anonymity and/or private Groups, the platforms removed the stigma, enabling like-minded people, including extremists, to find one another, communicate, and, eventually, to lose the fear of social stigma.

On the internet, even the most socially unacceptable ideas can find an outlet. As a proponent of free speech, I believe every person is entitled to speak his or her mind. Unfortunately, anonymity, the ability to form Groups in private, and the hands-off attitude of platforms have altered the normal balance of free speech, often giving an advantage to extreme voices over reasonable ones. In the absence of limits imposed by the platform, hate speech, for example, can become contagious. The fact that there are no viable alternatives to Facebook and Google in their respective markets places a special burden on those platforms with respect to content moderation. They have an obligation to address the unique free-speech challenges posed by their scale and monopoly position. It is a hard problem to solve, made harder by continuing efforts to deflect responsibility. The platforms have also muddied the waters by frequently using free-speech arguments as a defense against attacks on their business practices.

Whether by design or by accident, platforms empower extreme views in a variety of ways. The ease with which like-minded extremists can find one another creates the illusion of legitimacy. Protected from real-world stigma, communication among extreme voices over internet platforms generally evolves to more dangerous language. Normalization lowers a barrier for the curious; algorithmic reinforcement leads some users to increasingly extreme positions. Recommendation engines can and do exploit that. For example, former YouTube algorithm engineer Guillaume Chaslot created a program to take snapshots of what YouTube would recommend to users. He learned that when a user watches a regular 9/11 news video, YouTube will then recommend 9/11 conspiracies; if a teenage girl watches a video on food dietary habits, YouTube will recommend videos that promote anorexia-related behaviors. It is not for nothing that the industry jokes about YouTube's "three degrees of Alex Jones," referring to the notion that no matter where you start, YouTube's algorithms will often surface a Jones conspiracy theory video within three recommendations. In an op-ed in *Wired*, my colleague Renée DiResta quoted YouTube chief product officer Neal Mohan as saying that 70 percent of the views on his platform are from recommendations. In the absence of a commitment to civic responsibility, the recommendation engine will be programmed to do the things that generate the most profit. Conspiracy theories cause users to spend more time on the site.

Once a person identifies with an extreme position on an internet platform, he or she will be subject to both filter bubbles and human nature. A steady flow of ideas that confirm beliefs will lead many users to make choices that exclude other ideas both online and off. As I learned from Clint Watts, a national security consultant for the FBI, the self-imposed blocking of ideas is called a preference bubble. Filter bubbles are imposed by others, while a preference bubble is a choice. By definition, a preference bubble takes users to a bad place, and they may not even be conscious of the change.

Preference bubbles can be all-encompassing, especially if a platform like Facebook or Google amplifies them with a steady diet of reinforcing content. Like filter bubbles, preference bubbles increase time on site, which is a driver of revenue. In a preference bubble, users create an alternative reality, built around values shared with a tribe, which can focus on politics, religion, or something else. They stop interacting with people with whom they disagree, reinforcing the power of the bubble. They go to war against any threat to their bubble, which for some users means going to war against democracy and legal norms. They disregard expertise in favor of voices from their tribe. They refuse to accept uncomfortable facts, even ones that are incontrovertible. This is how a large minority of Americans abandoned newspapers in favor of talk radio and websites that peddle conspiracy theories. Filter bubbles and preference bubbles undermine democracy by eliminating the last vestiges of common ground among a huge percentage of Americans. The tribe is all that matters, and anything that advances the tribe is legitimate. You see this effect today among people whose embrace of

Donald Trump has required them to abandon beliefs they held deeply only a few years earlier. Once again, this is a problem that internet platforms did not invent. Existing fissures in society created a business opportunity that platforms exploited. They created a feedback loop that reinforces and amplifies ideas with a speed and at a scale that are unprecedented.

In his book, *Messing with the Enemy*, Clint Watts makes the case that in a preference bubble, facts and expertise can be the core of a hostile system, an enemy that must be defeated. As Watts wrote, "Whoever gets the most likes is in charge; whoever gets the most shares is an expert. Preference bubbles, once they've destroyed the core, seek to use their preference to create a core more to their liking, specially selecting information, sources, and experts that support their preferred alternative reality rather than the real, physical world." The shared values that form the foundation of our democracy proved to be powerless against the preference bubbles that have evolved over the past decade. Facebook does not create preference bubbles, but it is the ideal incubator for them. The algorithms ensure that users who like one piece of disinformation will be fed more disinformation. Fed enough disinformation, users will eventually wind up first in a filter bubble and then in a preference bubble. If you are a bad actor and you want to manipulate people in a preference bubble, all you have to do is infiltrate the tribe, deploy the appropriate dog whistles, and you are good to go. That is what the Russians did in 2016 and what many are doing now.

THE SAD TRUTH IS that Facebook and the other platforms are real-time systems with powerful tools optimized for behavior modification. As users, we sometimes adopt an idea suggested by the platform or by other users on the platform as our own. For example, if I am active in a Facebook Group associated with a conspiracy theory and then stop using the platform for a time, Facebook will do something surprising when I return. It may suggest other conspiracy theory Groups to join because they share members with the first conspiracy Group. And because conspiracy theory Groups are highly engaging, they are very likely to encourage reengagement with the platform. If you join the Group, the choice appears to be yours, but the reality is that Facebook planted the seed. It does so not because conspiracy theories are good for you but because conspiracy theories are good for them.

Research suggests that people who accept one conspiracy theory have a high likelihood of accepting a second one. The same is true of inflammatory disinformation. None of this was known to me when I joined forces with Tristan. In combination with the events I had observed in 2016, Tristan's insights jolted me, forcing me to accept the fact that Facebook, YouTube, and Twitter had created systems that modify user behavior. They should have realized that global scale would have an impact on the way people used their products and would raise the stakes for society. They should have anticipated violations of their terms of service and taken steps to prevent them. Once made aware of the interference, they should have cooperated with investigators. I could no longer pretend that Facebook was a victim. I cannot overstate my disappointment. The situation was much worse than I realized.

The people at Facebook live in their own preference bubble. Convinced of the nobility of their mission, Zuck and his employees reject criticism. They respond to every problem with the same approach that created the problem in the first place: more AI, more code, more short-term fixes. They do not do this because they are bad people. They do this because success has warped their perception of reality. To them, connecting 2.4 billion people is so obviously a good thing, and continued growth so important, that they cannot imagine that the problems that have resulted could be in any way linked to their designs or business decisions. It would never occur to them to listen to critics—how many billion people have the critics connected?—much less to reconsider the way they do business. As a result, when confronted with evidence that disinformation and fake news spread over Facebook influenced the Brexit referendum in the United Kingdom and a presidential election in the United States, Facebook took steps that spoke volumes about the company's world view. They demoted publishers in favor of family, friends, and Groups on the theory that information from those sources would be more trustworthy. The problem is that family, friends, and Groups are the foundational elements of filter and preference bubbles. Whether by design or by accident, they share the very disinformation and fake news that Facebook should want to suppress.

AS TRISTAN DESCRIBES IT, and Fogg teaches it, there are ten tools that platforms use to manipulate the choices of their users. Some of these tools relate to the interface design of each platform: menus, news feeds, and notifications. Platforms like Facebook would have you believe that the user is always in control, but as I have said before, user control is an illusion. Maintaining that illusion is central to every platform's success, but with Facebook, it is especially disingenuous. Menu choices limit user actions to things that serve Facebook's interests. In addition, Facebook's design teams exploit what are known as "dark patterns" in order to produce desired outcomes. Wikipedia defines a dark pattern as "a user interface that has been carefully crafted to trick users into doing things." The company tests every pixel to ensure it produces the desired response. Which shade of red best leads people to check their notifications? For how many milliseconds should notifications bubbles appear in the bottom left before fading away, to most effectively keep users on site? Based on what measures of closeness should we recommend new friends for you to "add"?

When you have more than two billion users, you can test every possible configuration. The cost is small. It is no accident that Facebook's terms of service and privacy settings, like those for most internet platforms, are hard to find and nearly impossible to understand. Facebook places a button on the landing page to give users access to the terms of service, but few people click on it. The button is positioned so that hardly anyone even sees it. Those who do see the button have been trained since the early days of the internet to believe that terms of service are long and incomprehensible, so they don't press it either. Facebook's terms of service have one goal and one goal only: to protect the company from legal liability. By using the platform, we give Facebook permission to do just about anything it wants.

Another tool from the Fogg tool kit is the "bottomless bowl." News Feeds on Facebook and other platforms are endless. In movies and television, scrolling credits signal to the audience that it is time to move on, providing what Tristan would call a "stopping cue." Platforms with endless news feeds and autoplay remove that signal, ensuring that users maximize their time on site for every visit. Endless news feeds work on dating apps. They work on photo sites like Instagram. And they work on Facebook. YouTube, Netflix, and Facebook use autoplay on their videos because it works, too. Next thing you know, millions of people are sleep deprived from binging on videos, checking Instagram, or browsing Facebook.

Notifications are another way that platforms exploit the weakest elements of human psychology. Notifications exploit an old sales technique, called the "foot in the door" strategy, that lures the prospect with an action that appears to be low cost but sets in motion a process that leads to bigger costs. Who wouldn't want to know they have just received an email, text, friend request, or Like? As humans, we are not good at forecasting the true cost of engaging with a foot-in-the-door strategy. Worse yet, we behave as though notifications are personal to us, completely missing that they are automatically generated, often by an algorithm tied to an artificial intelligence that has concluded that the notification is just the thing to provoke an action that will serve the platform's economic interests. That is not even the worst thing about notifications. I will get to that in a moment.

The persuasive technology tricks espoused by Fogg include several related to social psychology: a need for approval, a desire for reciprocity, and a fear of missing out. Everyone wants to feel approved of by others. We want our posts to be liked. We want people to respond to our texts, emails, tags, and shares. The need for social approval is what made Facebook's Like button so powerful. By controlling how often a user experiences social approval, as evaluated by others, Facebook can get that user to do things that generate billions of dollars in economic value. This makes sense because the currency of Facebook is attention. Users manicure their image in the hope of impressing others, but they soon discover that the best way to get attention is through emotion and conflict. Want attention online? Say something outrageous. This phenomenon first emerged decades ago in online forums such as The WELL, which often devolved into mean-spirited confrontation, and has reappeared in every generation of tech platform since then.

Social approval has a twin: social reciprocity. When we do something for someone else, we expect them to respond in kind. Likewise, when a person does something for us, we feel obligated to reciprocate. When someone "follows" us on Instagram, we feel obligated to "follow" them in return. When we see an "Invitation to Connect" on LinkedIn from a friend, we may feel guilty if we do not reciprocate the gesture and accept it. It feels organic, but it is not. Millions of users reciprocate one another's Likes and friend requests all day long, not aware that platforms orchestrate all of this behavior upstream, like a puppet master. As I noted in chapter 3, one of the most manipulative reciprocity tricks played by Facebook relates to photo tagging. When users post a photo, Facebook offers an opportunity to tag friends—the message "[Friend] has tagged you in a photo" is an appealing form of validation—which initiates a cycle of reciprocity, with notifications to users who have been tagged and an invitation to tag other people in the photo. Tagging was a game changer for Facebook because photos are one of the main reasons users visit Facebook daily in the first place. Each tagged photo brings with it a huge trove of data and metadata about location, activity, and friends, all of which can be used to target ads more effectively. Thanks to photo tagging, users have built a giant database of photos for Facebook, complete with all the information necessary to monetize it effectively. Other platforms play this game, too, but not at Facebook's scale. For example, Snapchat offers Streaks, a feature that tracks the number of consecutive days a user has traded messages with each person in his or her contacts list. As they build and the number of them grows, Streaks take on a life of their own. For the teens who dominate Snapchat's user base, Streaks can soon come to embody the essence of a relationship, substituting a Streak number for the elements of true friendship.

Another emotional trigger is fear of missing out (FOMO), which drives users to check their smartphone every free moment, as well as at times when they have no business doing so, such as while driving. FOMO makes notifications enticing. Have you ever tried to deactivate your Facebook account? As the software developer and blogger Matt Refghi discovered, Facebook will present a confirmation screen that shows the faces of several of your closest friends with text underneath: "[Friend] will miss you." For teens on Instagram or Snapchat, FOMO and the need for social approval combine to magnify the already stressful social lives of teenagers. Teenagers are particularly vulnerable to social pressure, and internet platforms add complexity to that equation that we are only beginning to understand.

The business choices of internet platforms compound the harm of persuasive technologies. Platforms work very hard to grow their user count but operate with little regard for users as individuals. The customer service department is reserved for advertisers. Users are just the fuel, so there is no one for them to call. The level of automation in the platforms is so great that when things go wrong operationally—an account gets hacked or locked, or the platform mistakenly blocks the user—the user generally must jump through hoops to rectify the problem. When a platform changes its terms of service, there is generally no obvious disclosure. Usage of the platform translates into acceptance of the terms of service. In short, users have no say in the relationship. In the event of a disagreement, their only option is to discontinue use of the service and lose access to an entire network of communication, which may include social opportunities and work opportunities. For services as ubiquitous as Facebook and Google, that is not a reasonable choice, which is why the platforms require it. They know that human psychology—the desire to protect an investment of time and content—and network effects will keep users on board, no matter how poorly the platforms treat them. As a condition of the terms of service, legal disagreements are subject to arbitration rather than litigation, which favors the platform. And for Facebook, at least, there truly is no alternative. No other platform can reproduce the functionality, much less the scale, of Facebook. The company has monopoly power.

NONE OF US WANTS to admit to any addiction, but millions have texted while driving. We like to think of ourselves as being in control. As a lifelong technology optimist and an early adopter of new products, I was at far-greater-than-average risk of tech addiction. For example, I got an iPhone on the first day they were available and have bought into each subsequent generation on day one. I am a compulsive checker of my phone, despite having turned off notifications and gotten rid of several apps. I did not understand my behavioral addiction until I joined forces with Tristan in April 2017. Until then, I thought I bore sole responsibility for the problem. I assumed that only people with my heat-seeking love for technology could fall victim to unhelpful tech behaviors. Tristan opened my eyes to the reality that technology companies had devoted some of their best minds to exploiting the weaknesses in human psychology. They did so on purpose. To make money. And when they had made themselves ridiculously wealthy, they kept doing it because it never occurred to them to do anything else. When called to account for this, tech companies blame pressure from shareholders. Given that the founders of both Facebook and Google have total control of their companies, that excuse falls short.

Few of us can resist the lure of persuasive technology. All we can do is minimize the stimulus, or avoid it altogether by not using devices. Each element of persuasive technology is a way to fool the user. Tristan knew many of the concepts from magic, but it was in Fogg's class that he came to appreciate that when ported to a computer or smartphone, those concepts gave software the power to monopolize attention. He began to understand that "engagement" was just a play for users' attention . . . and time. Pushed hard enough, these tricks can strip users of agency.

THE PROBLEMS WITH INTERNET platforms on smartphones extend far beyond addiction. They also pollute the public square by empowering negative voices at the expense of positive ones. From its earliest days, the internet's culture advocated free speech and anonymity without constraint. At small scale, such freedom of speech was liberating, but the architects of the World Wide Web failed to anticipate that many users would not respect the culture and norms of the early internet. At global scale, the dynamics have changed to the detriment of civil discourse. Bullies and bad actors take advantage. The platforms have made little effort to protect users from harassment, presumably due to some combination of libertarian values and a reluctance to enforce restrictions that might reduce engagement and economic value. They have not created internal systems to limit the damage from bad actors. They have not built in circuit breakers to limit the spread of hate speech. What they do instead is ban hate speech and harassment in their terms of service, covering their legal liability, and then apologize when innocent users suffer harm. Twitter, Facebook, and Instagram all have a bully problem, each reflecting the unique architecture and culture of the platform. The interplay of platforms also favors bad actors. They can incubate pranks, conspiracy theories, and disinformation in fringe sites like 4chan, 8chan, and Reddit, which are home to some of the most extreme voices on the internet, jump to Twitter to engage the press, and then, if successful, migrate to Facebook for maximum impact. The slavish tracking of Twitter by journalists, in combination with their willingness to report on things that trend there, has made news organizations complicit in the degradation of civil discourse.

IN AN ESSAY in the *MIT Technology Review*, UNC professor Zeynep Tufekci explained why the impact of internet platforms on public discourse is so damaging and hard to fix. "The problem is that when we encounter opposing views in the age and context of social media, it's not like reading them in a newspaper while sitting alone. It's like hearing them from the opposing team while sitting with our fellow fans in a football stadium. Online, we're connected with our communities, and we seek approval from our like-minded peers. We bond with our team by yelling at the fans of the other one. In sociology terms, we strengthen our feeling of 'in-group' belonging by increasing our distance from and tension with the 'out-group'—us versus them. Our cognitive universe isn't an echo chamber, but our social one is. This is why the various projects for fact-checking claims in the news, while valuable, don't convince people. Belonging is stronger than facts."

Facebook's scale presents unique challenges for democracy. Zuck's vision of connecting the world and bringing it together may be laudable in intent, but the company's execution leaves much to be desired. Prohibiting hate speech in the terms of service provides little sanctuary for users. "Community standards" vary from country to country, most often favoring the powerful over the powerless. The company needs to learn how to identify emotional contagion and contain it before there is significant harm. It should also face an uncomfortable truth: if it wants to be viewed as a socially responsible company, it may have to abandon its current policy of openness to all voices, no matter how damaging. Being socially responsible may also require the company to compromise its growth targets.

Now that we know that Facebook has a huge influence on our democracy, what are we going to do about it? For all intents and purposes, we have allowed these people to have a profound influence on the course for our country and the world with no input from the outside.

THERE ARE SO MANY obvious benefits to social networks that we find it hard to accept that the platforms have poisoned political discourse in democracies around the world. The first thing to understand is that the problem is not social networking per se, but a set of choices made by entrepreneurs to monetize it. To make their advertising business model work, internet platforms like Facebook and YouTube have inverted the traditional relationship of technology to humans. Instead of technology being a tool in service to humanity, it is humans who are in service to technology. We are the fuel.

Tristan took Fogg's class a few months after my first meeting with Zuck. The challenge for would-be entrepreneurs at that time was

Google, already a dominant company worth more than one hundred billion dollars. Google derived almost all its profit from advertising on its search engine. When users searched for products, Google helpfully served up an ad relevant to that search. Google focused maniacally on shrinking the amount of time necessary to find what the user desired, with keyword-based ads in the sidebar. In short, Google's advertising model did not depend on maximizing attention. At least, not then. (That would change a few years later, when the company began to monetize YouTube.)

There was no way to compete with Google in search, so aspiring entrepreneurs in the first decade of the new millennium looked for opportunities outside Google's areas of dominance. The idea of Web 2.0, where the focus shifted from pages to people, was gaining traction, and "social" emerged as a buzzword. LinkedIn launched its social network for business in 2003, the year before Zuck launched Facebook. In the early days, attracting new users dominated the strategy of both companies, pushing all other factors, including attention, temporarily to the sideline. LinkedIn and Facebook quickly emerged as winners, but neither was so dominant as to preclude new niche players.

As noted in chapter 2, the technology entrepreneurs who gave birth to social media enjoyed the benefits of perfect timing. They did not face the constraints on processing power, memory, storage and bandwidth that had defined the first fifty years of Silicon Valley. Simultaneously, there was an unprecedented surplus in venture capital funding. As we've seen, it had never been less costly to launch a startup, and the consumer-facing opportunities enabled by software stacks, cloud computing, and pervasive 4G wireless were larger than the industry had ever seen. Entrepreneurs could hire cheaper, less experienced engineers and mold them into organizations that pursued target metrics single-mindedly. At a time when technology could do practically anything, entrepreneurs chose to exploit weaknesses in human psychology. We're just beginning to understand the implications of that.

Freed from traditional engineering constraints, entrepreneurs worked to eliminate friction of other kinds, starting with the retail price. Consumers are often willing to try a free product that they would never touch if they had to pay first. Every social product was free to use, though some offered users in-app purchases to generate revenue. Everyone else monetized with advertising. Other forms of friction, such as regulation and criticism, could be overcome with a combination of promises, apologies, and a refusal to permit outsiders to monitor compliance. Eliminating friction enabled the platforms to compete for attention. They competed against other products. They competed against other leisure and work activities. And as Netflix CEO Reed Hastings memorably observed, they competed against sleep.

In the early years of social media, the default internet platform—a web browser on a personal computer—was physically awkward, requiring a desk and somewhere to sit. On both desktops and notebooks, the battle for attention favored web-native applications over older media, such as news, television, books, and film. With rapid adaptation as the engineering philosophy of the moment, web-native apps exploited every opportunity to increase their share of attention. One of the most important vectors for innovation was personalization, with the goal of delivering a unique experience for every user, where traditional media offered the same content to everyone. Stuck with inflexible product designs and an inability to innovate rapidly, traditional media could not keep pace.

When the iPhone shipped in 2007, it looked like no mobile phone before it—a thin, flat rectangle with rounded corners. The iPhone featured a virtual keyboard, generally thought at that time to be an unserious approach that would not catch on in the business market, where BlackBerry ruled with physical keyboards. The iPhone's main functions were phone, email, music, and web. Users went crazy. So compelling was the experience of using an iPhone that it altered the relationship between device and human. In the 3G era, Wi-Fi enabled smartphones to deliver compelling web media experiences every waking moment of the day. Opportunities for persuasion grew geometrically, especially when Apple opened the App Store for iPhone applications on July 10, 2008. Facebook, Twitter, and other social platforms took advantage of the App Store to accelerate their migration to mobile. Within a few years, mobile would dominate the social media industry.

Competing for attention does not sound like a bad thing. Media companies have done it since the launch of the *New York Sun* in 1833, at the very least. Parents have long complained about kids watching too much television, listening to music night and day, or spending too much time playing video games. Why should we be concerned now? If you create a timeline of concerns about kids and media, it might start with comic books, television, and rock 'n' roll music in the fifties. Since then, technology has evolved from slow and artificial to real-time and hyperrealistic. Products today use every psychological trick to gain and hold user attention, and kids are particularly vulnerable. Not surprisingly, medical diagnoses related to kids' use of technology have exploded. Internet platforms, video games, and texting present different kinds of problems, but all three are much more immersive than analogous products twenty years ago. People of all ages spend a huge portion of their waking hours on technology platforms. In his book *Glow Kids*, Nicholas Kardaras cited a 2010 study by the Kaiser Family Foundation that said children between the ages of eight and eighteen spend nine and a half hours a day on screens and phones. Seven and a half of those hours are spent on television, computers, and game consoles, with an additional ninety minutes for texting and thirty minutes otherwise on the phone.

Overconsumption of media is not a new problem, but social apps on smartphones have taken the consequences to new levels. The convenience and compelling experience of smartphones enabled app developers—certainly those trained by B. J. Fogg—to create products that mimic the addiction-causing attributes of slot machines and video games. Apps created by Fogg's students were particularly adept at monopolizing user attention. Nearly all of Fogg's students embraced that mission. Tristan Harris did not.

Tristan left Stanford's master's program in computer science to launch a startup called Apture after he completed the term in Fogg's class. His idea for Apture was to enhance text-based news with relevant multimedia, specifically video that explained concepts in the news story to which it was tied. My friend Steve Vassallo, a venture capitalist who had mentored Tristan in a program at Stanford, asked me to take a meeting with Apture's founders in the spring of 2007. After introducing Tristan to the team at Forbes, I lost touch. During that time, Apture found customers but never took off. Google acquired the company in late 2011. The deal was what Silicon Valley calls an "acqui-hire," where the acquiring company pays enough to cover the debts and possibly a return of capital to investors in exchange for a complete team of engineers. All the Apture team got out of the deal were jobs at Google, but the acquisition set Tristan on a new course, one that would bring us back together in 2017.

Shortly after arriving at Google, Tristan had an epiphany: a battle for user attention would not be good for users. Tristan created a slide deck to share his concerns, and the deck went viral inside Google. The company made no commitment to implementing the ideas, but it rewarded Tristan with an opportunity to design his own job. He picked New York City and created a new position: design ethicist. From that moment, Tristan became an evangelist for humane design principles. In Tristan's mind, the design of technology products should prioritize the users' well-being. For any given product, there are design choices that can either be humane or not. Monochrome smartphone displays are more humane than brightly colored displays, as they trigger less dopamine. A humane design would reduce the number of notifications and package them in ways that are respectful of human attention. There is a lot that could be done to improve today's smartphones and internet platforms with humane design principles, and some vendors have begun to take steps. Humane design focuses on reducing the addictive power of technology, which is really important but is not a complete answer, even for Tristan. It is part of a larger philosophical approach, human-driven technology, which advocates returning technology to the role of being a tool to serve user needs rather than one that exploits users and makes them less capable. With its focus on user interface issues, humane design is a subset of human-driven technology, which also incorporates things like privacy, data security, and applications functionality. We used to take human-driven technology for granted. It was the philosophical foundation for Steve Jobs's description of computers as a "bicycle for the mind," a tool that creates value through exercise as well as fun. Jobs thought computers should make humans more capable, not displace or exploit them. Every successful tech product used to fit the Jobs model, and many still do. Personal computers still empower the workers who use them. There should be a version of social media that is human-driven. Facebook and Google seem to have discarded the bicycle for the mind metaphor. They advocate AI as a replacement for human activity, as described in a recent television ad for Google Home: "Make Google do it."

At Google, Tristan saw that users effectively served artificial intelligence, rather than the other way around. Tristan decided to change that and found a handful of supporters within Google and more on the outside. At Google, Tristan's principal ally was an engineer named Joe Edelman. From their collaboration grew a website and movement called Time Well Spent, which they launched in 2013. Time Well Spent offered advice on managing time in a world filled with distractions, while also advocating for human-centered design. It grew steadily to sixty thousand members, many of whom were technology-industry people troubled by attention-sucking platforms. Time Well Spent attracted true believers, but it struggled to effect change.

In 2016, a creative and well-connected Irishman named Paddy

Cosgrave invited Tristan to speak at his Web Summit conference in Lisbon. Paddy had built a personal brand and business by bringing young tech entrepreneurs, investors, and wannabes together at a string of increasingly high-profile conferences that emphasized personal networking. Tristan gave a well-received presentation about how internet platforms on smartphones hacked the brains of their users. He described the public health consequences of brain hacking as a loss of personal agency, a loss of humanity. Among the attendees was a producer from *60 Minutes* named Andy Bast, in Lisbon to scout for new stories. Bast liked what he heard and offered Tristan an opportunity to appear on the show.

Tristan's *60 Minutes* segment aired on April 9, 2017. Three days later, we joined forces. While Tristan was well-connected within his peer group in tech, he had few relationships outside the industry. I had once had a large network inside and outside tech, but I had cut back professionally and lost touch with much of it. When we thought about possible influencers for a national conversation, we realized we knew only a few people we could contact, all in technology and media. We had no relationships in government.

The first potential opportunity was only a couple of weeks away, at the annual TED Conference in Vancouver, British Columbia, ground zero for TED Talks. It would be the perfect platform for sharing Tristan's message to leaders from the technology and entertainment industries, but we did not know if the organizers were even aware of Tristan's ideas. They certainly had not offered an invitation to speak. Then a miracle occurred. Eli Pariser, whose legendary presentation on filter bubbles had mesmerized the TED audience in 2011, independently suggested to TED curator Chris Anderson that he add Tristan to the program. And so it happened at the last minute.

Presenters at the TED Conference usually spend six months preparing for their eighteen minutes onstage. Tristan had barely more than a week. Using no slides, he rose to the occasion and delivered a powerful talk. We were hopeful that the TED audience would embrace the ideas and offer to help. What we got instead was polite interest but little in the way of follow-up.

We should not have been surprised. Facebook, Google, Twitter, LinkedIn, Instagram, Snapchat, WhatsApp, and the other social players have created more than one trillion dollars in wealth for their executives, employees, and investors, many of whom attend TED. To paraphrase Upton Sinclair, it is difficult to get a person to embrace an idea when his or her net worth depends on not embracing it.

We took two insights away from TED: we needed to look outside tech for allies, and we needed to consider other formulations of our message besides brain hacking. If we were to have any hope of engaging a large audience, we needed to frame our argument in ways that would resonate with people outside Silicon Valley.

5

Mr. Harris and Mr. McNamee
Go to Washington

Every aspect of human technology has a dark side,
including the bow and arrow. —MARGARET ATWOOD

A few weeks after TED, a friend passed me the contact information for an aide to Senator Mark Warner, co-chair of the Senate Intelligence Committee. I called the aide, explained what we were doing, and asked a question: "Who is going to prevent the use of social media to interfere in the 2018 and 2020 elections?" Senate Intelligence knew about the role that social media had played in the Russian interference, but the committee had oversight responsibility for intelligence matters, not social media. Their focus was on the hacks of the DNC and DCCC, as well as the meeting in Trump Tower of senior members of the Trump campaign with Russian agents. They would not normally investigate things that happened inside the servers of Facebook. But the aide recognized that Senate Intelligence might be the only committee with the ability to investigate the threat from social media and agreed to arrange a meeting with the senator. It took a couple of months, but in July 2017, we went to Washington. Then things got more interesting.

There is a popular misconception that regulation does not work with technology. The argument consists of a set of flawed premises: (1) regulation cannot keep pace with rapidly evolving technology, (2) government intervention always harms innovation, (3) regulators can never understand tech well enough to provide oversight, and (4) the market will always allocate resources best. The source of the misconception is a very effective lobbying campaign, led by Google, with an assist from Facebook. Prior to 2008, the tech industry maintained an especially low profile in Washington. All of that changed when Google, led by chairman Eric Schmidt, played a major role in Barack Obama's first presidential campaign, as did Facebook cofounder Chris Hughes. Obama's win led to a revolving door between Silicon Valley and the executive branch, with Google dominating the flow in both directions. The Obama administration embraced technology, and with it the optimism so deeply ingrained in Silicon Valley. Over the course of eight years, the relationship between Silicon Valley and the federal government settled into a comfortable equilibrium. Tech companies supported politicians with campaign contributions and technology in exchange for being left alone. The enormous popularity of tech companies with voters made the hands-off policy a no-brainer for members of Congress. A few commentators expressed concern about the cozy relationship between tech and Washington, some out of concern about tech's outsized influence and market power and others out of fear that the internet platforms might not be the neutral parties to democracy they claimed to be.

Once you get past the buzzwords, tech is not particularly complicated in comparison to other industries that Congress regulates. Health care and banking are complex industries that Congress has been able to regulate effectively, despite the fact that relatively few policy makers have had as much involvement with them as they have with technology, which touches everyone, including members of Congress, on a daily basis. The decision about whether or not to impose new regulations on tech should depend on a judgment by the policy makers that the market has failed to maintain a balance among the interests of industry, customers, suppliers, competitors, and the country as a whole. Critics who charge that regulation is too blunt an instrument for an industry like tech are not wrong, but they miss the point. The goal of regulation is to change incentives. Industries that ignore political pressure for reform, as the internet platforms have, should expect ever more onerous regulatory initiatives until they cooperate. The best way for tech to avoid heavy regulation is for the industry leaders to embrace light regulation and make appropriate changes to their business practices.

In July 2017, when Tristan and I arrived in Washington, DC, the town remained comfortably in the embrace of the major tech platforms. Google and Facebook led a large tech industry presence on Capitol Hill. Facebook director Peter Thiel continued to advise President Trump, leading high-profile meetings in the White House with technology executives. We had been able to secure four meetings: with a commissioner of the Federal Trade Commission (FTC), with executives from a leading think tank focused on antitrust policy, and with two senators, Elizabeth Warren and Mark Warner.

The FTC has two important mandates: consumer protection and prevention of anticompetitive business practices. Created in 1914, the FTC plays an essential role in balancing the interests of businesses, their customers, and the public, but deregulation and budget cuts have undermined the FTC over the past three decades. Our visit with Commissioner Terrell McSweeny came at a time when the FTC was effectively paralyzed. Only two of the five slots for commissioners were filled. No company in Silicon Valley worried about regulation from the FTC.

Commissioner McSweeny shared her perspective on how we might best approach the FTC when all the slots were filled, which finally took place in May 2018. She explained the FTC's consumer protection mandate and suggested that violations of the terms of service were the lowest hanging fruit for regulating software businesses.

The focus of our trip, however, was our meeting with Senator Mark Warner. We started with a briefing on our work as it related to public health and antimonopoly. Then I asked a variant of the question I had asked the senator's aide in May: Could Congress do something to prevent the use of social media to interfere in future elections? Senator Warner asked us to share our thoughts on what had happened during the 2016 election.

Tristan and I are not investigators. We didn't have evidence. What we had were hypotheses that explained what we thought must have happened. Our first hypothesis was that Russia had done much more than break into servers at the DNC and the DCCC and post some of what they took on WikiLeaks. There were too many other Russian connections swirling around the election for the hacks to be the whole story. For example, in 2014, a man named Louis Marinelli launched an effort to have California secede from the United States. He initially called the effort Sovereign California, built a presence on Facebook and Twitter to promote his cause, and then released a 165-page report with that name in 2015. The secession idea attracted support from some powerful Californians, including at least one very prominent venture capitalist. However, the secession movement did not begin with Californians. Marinelli, who was born in Buffalo, New York, had deep ties to Russia and lived there part-time while launching Sovereign California. At least some of his funding appears to have come from Russia, and Russian bots supported the effort on social media. In speaking with Senator Warner, we noted that there was also a Texas secession movement, and we hypothesized that it might have had Russian connections as well. These sites were committed to polarizing Americans over social media. What else had the Russians done?

What if Russia had run an entire campaign of discord and disinformation over social media? What would the goals have been of a Russian social media campaign? We hypothesized that the campaign must have started no later than 2014, when the California secession movement began. It might have started as early as 2013, when Facebook introduced Lookalike Audiences, the tool that allows advertisers to target every user on Facebook who shares a given set of characteristics. Given the way political discourse works on Facebook, Lookalike Audiences would have been particularly effective for targeting both true believers and those who could be discouraged from voting. Since Facebook provides advertising support services to all serious advertisers, we could not exclude the possibility that they may have helped the Russians execute their mischief. If the Russians did not use Lookalike Audiences, they missed a huge point of leverage.

In 2014, the Russian social media campaign probably would not have been focused on a candidate. We hypothesized that it focused instead on a handful of particularly divisive issues: immigration, guns, and perhaps conspiracy theories. We had noted an explosion of divisive content on Facebook and other social media platforms in the years before the 2016 election and hypothesized that Russia may have played a role.

How did the Russian agents figure out which Americans to target? Did they build a database of users organically, or did they acquire a database? If the latter, where did it come from? We did not know the answers, but we thought both were possible. The Russians had enough time to build audiences, particularly if they were willing to invest advertising dollars, which would have been most effective if used to build Facebook Groups. Groups have a couple of features that make them vulnerable to manipulation. Anyone can start a Group, and there is no guarantee that the organizer is the person he or she claims to be. In addition, there are few limits on the names of Groups, which enables bad actors to create Groups that appear to be more legitimate than they really are. I hypothesized to Senator Warner that some of the pro–Bernie Sanders Groups I encountered in early 2016 may have been part of the Russian campaign.

We hypothesized that the Russians might have seeded Facebook Groups across a range of divisive issues, possibly including Groups on opposite sides of the issue to maximize impact. The way the Groups might have worked is that a troll account—a Russian impersonating an American—might have formed the Group and seeded it with a number of bots. Then they could advertise on Facebook to recruit members. The members would mostly be Americans who had no idea they were joining a Group created by Russians. They would be attracted to an idea—whether it was guns or immigration or whatever—and once in the Group, they would be exposed to a steady flow of posts designed to provoke outrage or fear. For those who engaged frequently with the Group, the effect would be to make beliefs more rigid and more extreme. The Group would create a filter bubble, where the troll, the bots, and the other members would coalesce around an idea floated by the troll.

We also shared a hypothesis that the lack of data dumps from the DCCC hack meant the data might have been used in congressional campaigns instead. The WikiLeaks email dumps had all come from the DNC hack. The DCCC, by contrast, would have had data that might be used for social media targeting, but more significantly, Democratic Party data from every congressional district. The data would have been the equivalent of inside information about Democratic voters. Hypothetically, the DCCC data might have allowed the Russians—or potentially someone in the Republican Party—to see which Democrats in a district could be persuaded to stay home instead of voting. We learned later that during the final months of the 2016 campaign, the Russians had concentrated their spending in the states and congressional districts that actually tipped the election.

It seemed likely to us that the Russian interference would have been most effective during the Republican primary. The Russians had been inflaming Groups and spreading disinformation about polarizing topics for more than a year when the candidates began running for president. All but one of the seventeen candidates ran on a relatively mainstream Republican platform. The seventeenth candidate, Donald Trump, uniquely benefited from the Russian interference because he alone campaigned on their themes: immigration, white nationalism, and populism. Whether by design or by accident, Trump's nomination almost certainly owed something to the Russian interference.

In the general election campaign against Hillary Clinton, Trump benefited significantly from the Russian interference on Facebook. We hypothesized that the Russians—and Trump—would have focused on activating a minority—Trump's base—while simultaneously suppressing the vote of the majority. Facebook's filter bubbles and Groups would have made that job relatively easy. We didn't know how much impact the Russians would have had. However, roughly four million Obama voters did not vote in 2016, nearly fifty-two times the vote gap in the states Clinton needed to win but didn't.

I concluded with an observation: the Russians might have used Facebook and other internet platforms to undermine democracy and influence a presidential election for roughly one hundred million dollars, or less than the price of a single F-35 fighter. It was just an educated guess on my part, based on an estimate of what eighty to one hundred hackers might cost for three or four years, along with a really large Facebook ad budget. In reality, the campaign may have cost less, but given the outcome, one hundred million dollars would have been a bargain. The Russians might have invented a new kind of warfare, one perfectly suited to a fading economic power looking to regain superpower status. Our country had built a Maginot Line—half the world's expenditures on defense, plus hardened data centers in government and finance—and it never occurred to anyone that a bad actor could ignore that and instead use an American internet platform to manipulate the minds of American voters.

Senator Warner and his staff understood immediately. Senator Warner asked, "Do we have any friends in Silicon Valley?"

We suggested that Apple could be a powerful ally, as it had no advertising-supported businesses and had made data privacy a core feature of the Apple brand.

The senator brightened and asked, "What should we do?"

Tristan didn't miss a beat. "Hold a hearing and make Mark Zuckerberg testify under oath. Make him justify profiting from filter bubbles, brain hacking, and election interference."

Washington needed help dealing with the threat of interference through social networks, and Senator Warner asked that we support him by providing insight into the technology.

Two weeks later, in August 2017, the editor-in-chief of *USA Today* asked me to write an op-ed. It was titled "I Invested Early in Google and Facebook. Now They Terrify Me." I had never before written an opinion piece for a newspaper. CNBC invited me to discuss the issue on *Squawk Alley* three times in the next two weeks, breaking the story more broadly. The op-ed and TV appearances came just as a flurry of news about Facebook seemed to confirm every hypothesis we had shared with Senator Warner.

Despite mounting evidence, Facebook continued to deny it had played a role in the Russian interference. They seemed to think that the press would quickly lose interest in the story and move on. It's easy to see why they felt that way. Facebook knows more about user attention than anyone. They assumed the election-interference story was no different from any of a dozen scandals the company had faced before. Even Tristan and I assumed that Facebook would eventually prevail. Before that happened, we hoped that we could make enough people aware of the dark side of internet platforms on smartphones that Facebook employees might feel pressure to reform the company.

Then, on September 6, 2017, Facebook's vice president of security, Alex Stamos, posted "An Update on Information Operations on Facebook." Despite the innocuous headline, the post began with a bombshell: Facebook had uncovered one hundred thousand dollars of Russian spending on three thousand ads from June 2015 through May 2017. The three thousand ads were connected to 470 accounts that Facebook labeled as "inauthentic." One hundred thousand dollars does not sound like much of an ad buy, but a month later the researcher Jonathan Albright provided context when he pointed out that posts from only six of the Russia-sponsored Groups on Facebook had been shared 340 million times. Groups are built around a shared interest. Most of those doing the sharing would have been Americans who trusted that the leaders of the Group were genuine.

There are millions of Groups on Facebook, for just about every organization, personality, politician, brand, sport, philosophy, and idea. The ones built around extreme ideas—disinformation, fake news, conspiracy theories, hate speech—become filter bubbles, reinforcing the shared value, intensifying emotional attachments to it. Thanks to Groups and filter bubbles, inflammatory posts can reach huge numbers of like-minded people on Facebook with only a little spending. The Russian interference on social media was exceptionally cost effective.

Our fears about interference in future elections were compounded by the refusal of the Trump administration and congressional Republicans to investigate. We had no contact with Special Counsel Robert Mueller's investigation, but press reports indicated that Mueller might be fired at any time. Journalists reported that Mueller had aligned with New York State attorney general Eric Schneiderman, in whose jurisdiction money laundering and other possible crimes might have occurred. Common Sense Media founder Jim Steyer suggested that I meet with Schneiderman in New York. Jim set it up, and I met the attorney general for dinner at Gabriel's restaurant on West Sixtieth Street in Manhattan.

When I gave him the short-form description of our work, Schneider-man asked me to work with an advisor to his office, Tim Wu, a professor at Columbia Law School. Tim is the person who coined the term "net neutrality." He wrote a seminal book, *The Attention Merchants*, about the evolution of advertising from tabloid newspapers to persuasive-technology platforms like Facebook. When I met Tim a few days later, he helped me understand the role of state attorneys general in the legal system and the kind of evidence that would be necessary to make a case. Over the ensuing six months, he organized a series of meetings with staff in Schneiderman's office, a truly impressive group of people. We did not have to explain internet platforms to them. Not only did the New York AG's office understand the internet, they had data scientists who could perform forensics. The AG's office had the skills and experience to handle the most complex cases. In time, we would furnish them with whistle-blowers, as well as insights. By April 2018, thirty-seven state attorneys general had begun investigations of Facebook.

6

Congress Gets Serious

Technological progress has merely provided us with more
efficient means for going backwards. —ALDOUS HUXLEY

In the month following our visit to Washington, journalists validated many of the hypotheses we had shared with Senator Warner. The Russians really had interfered by focusing on divisive issues during the primary and supporting Trump in the general election, while disparaging Clinton throughout. Democrats on Capitol Hill were digging into the issue, and some influential members of Congress wanted us to be part of it. Before the end of August, we got a call from Washington asking that we return. They invited us for three days, and they scheduled every moment. Just in time for the trip, Renée DiResta joined our team. Renée is one of the world's foremost experts on how conspiracy theories spread over the internet. In her job at that time, as director of research at New Knowledge, Renée helped companies protect themselves from disinformation, character assassination, and smear attacks, the kind of tactics the Russians used in 2016. New Knowledge also created Hamilton 68, the public dashboard that tracks Russian disinformation on

Twitter. Sponsored by the German Marshall Fund and introduced on August 2, 2017, Hamilton 68 enables anyone to track what pro-Kremlin Twitter accounts are discussing and promoting.

Renée is also director of policy of Data for Democracy, whose mission is "to be an inclusive community of data scientists and technologists to volunteer and collaborate on projects that make a positive impact on society." Renée's own focus is on analysis of efforts by bad actors to subvert democracy around the world. Unlike us, Renée was a pro in the world of election security. She and her colleagues had heard whispers of Russian interference efforts in 2015 but had struggled to get the authorities to take action.

The daughter of a research scientist, Renée got her first computer at age five or six. She does not remember a time before she had one. Raised in Yonkers, New York, Renée's other love was music. She learned to play piano as a child and played it "competitively" into her college years. Renée started coding at age nine and volunteered in a research lab at the Sloan-Kettering hospital in eighth grade. She worked on a project that looked for a correlation between music training and temporal reasoning. It was a small-scale study, but it married Renée's two great interests. She earned a degree in computer science at SUNY Stony Brook before going into government service in a technology operations role. Renée underplays this part of her résumé, focusing instead on her time in algorithmic trading on Wall Street. In that job, she observed the tricks that market participants use to outsmart one another, some of which are similar to the tools hackers could use to interfere in an election.

Our first meeting in Washington was with Senator Warner. "I'm on your team," he said, by way of an opening. The rest of the meeting focused on the senator's desire to hold a hearing with the top executives of Facebook, Google, and Twitter to question them about their role in the Russian interference in the 2016 election. The committee staff was negotiating with the internet platforms, in the hope of securing participation by the CEOs. No announcement would be made until that negotiation was complete.

Later that first day, we had the first of two meetings with Representative Adam Schiff of California, the ranking member of the House Permanent Select Committee on Intelligence (HPSCI). The committee had fractured along party lines, and Schiff had the unenviable task of attempting a traditional approach to congressional oversight, despite aggressive opposition by his committee's chair. The minority party has no power in the House of Representatives, which made the task exceptionally frustrating. Representative Schiff asked Renée and me to meet with the committee staff, which we did in a secure facility near the Capitol. The conversation turned to our hypotheses and what they implied for future elections. Renée characterized techniques the Russians may have used to spread disinformation on social media. The Russians' job was made easier by the thriving communities of libertarians and contrarians on the internet. They almost certainly focused on sites that promoted anonymous free speech, sites like Reddit, 4chan, and 8chan. These sites are populated by a range of people, but especially those who hold views that may not be welcome in traditional media. Some of these users are disaffected, looking for outlets for their rage. Others are looking to expose what they see as the hypocrisy of society. Still others have agendas or confrontational personalities looking for an outlet. And some just want to play pranks on the world, preying on the gullibility of internet users, just to see how far they can push something outrageous or ridiculous. These and other sites would have been fertile ground for the Russian messages on immigration, guns, and white nationalism. They were also ideal incubators for disinformation.

Renée explained that the typical path for disinformation or a conspiracy theory is to be incubated on sites like Reddit, 4chan, or 8chan. There are many such stories in play at any time, a handful of which attract enough support to go viral. For the Russians, any time a piece of disinformation gained traction, they would seed one or more websites with a document that appeared to be a legitimate news story about the topic. Then they would turn to Twitter, which has replaced the Associated Press as the news feed of record for journalists. The idea was to post the story simultaneously on an army of Twitter accounts, with a link to the fake news story. The Twitter army might consist of a mix of real accounts and bots.

If no journalist picked up the story, the Twitter accounts would post new messages saying some variant of "read the story that the mainstream media doesn't want you to know about." Journalism is intensely competitive, with a twenty-four-hour news cycle that allows almost no time for reflection. Eventually some legitimate journalist may write about the story. Once that happens, the game really begins. The army of Twitter accounts—which includes a huge number of bots—tweets and retweets the legitimate story, amplifying the signal dramatically. Once a story is trending, other news outlets are almost certain to pick it up. At that point, it's time to go for the mass market, which means Facebook. The Russians would have placed the story in the Facebook Groups they controlled, counting on Facebook's filter bubbles to ensure widespread acceptance of the veracity of the story, as well as widespread sharing. Trolls and bots help, but the most successful disinformation and conspiracy theories leveraged American citizens who trusted the content they received from fellow members of Facebook Groups.

An example is Pizzagate, a disinformation story that claimed that emails found by the FBI on the laptop of Anthony Weiner, a disgraced former member of Congress and the husband of the vice chair of Hillary Clinton's presidential campaign, suggested the presence of a pedophilia ring connected to members of the Democratic Party at pizza parlors in the Washington area. The theory went on to say that emails stolen from the Democratic National Committee included coded messages about pedophilia and human trafficking and that one pizza parlor in particular might be home to satanic rituals. The story, which appeared nine days before the 2016 election, was a complete fabrication, but many people believed it. One man bought in to such a degree that he showed up at the pizza parlor in December, armed with an AR-15, and fired three shots into the building. Fortunately, no one was hurt.

Untangling a conspiracy theory is tricky, but Pizzagate's origins are clearer than most. The story first appeared on a white supremacist Twitter account. Wikipedia states that accounts on 4chan and Twitter parsed the stolen DNC emails in search of coded messages and claimed to have found many. A range of conspiracy sites, including Infowars, picked up the story and amplified it on the far right. The story spread in the month after the election, with Twitter playing a large role. A subsequent analysis by the researcher Jonathan Albright indicated that a disproportionate share of the tweets came from accounts in the Czech Republic, Cyprus, and Vietnam, while most of the retweets were from bot accounts. The Czech Republic, Cyprus, and Vietnam? What was that about? Were they Russian agents or just enterprising entrepreneurs? The 2016 US presidential election seems to have attracted some of each. In a world where web traffic anywhere can be monetized with advertising and where millions of people are gullible, entrepreneurs will take advantage. There had been widespread coverage of young men in Macedonia whose efforts to sell ads against fabricated news stories mostly failed with Clinton voters but worked really well with fans of Trump and, to a lesser extent, Bernie Sanders.

On its face, the Pizzagate conspiracy theory is unbelievable. A pedophilia ring associated with the Democratic Party at a pizza parlor in DC? Coded messages in emails? And yet people believed it, one so deeply that he, in his own words, "self-investigated it" and fired three bullets into the pizza parlor. How can that happen? Filter bubbles. What differentiates filter bubbles from normal group activity is intellectual isolation. Filter bubbles exist wherever people are surrounded by people who share the same beliefs and where there is a way to keep out ideas that are inconsistent with those beliefs. They prey on trust and amplify it. They can happen on television when a channel blocks opposing viewpoints. Platforms have little incentive to eliminate filter bubbles because they improve metrics that matter: time on site, engagement, sharing. They can create the illusion of consensus where none exists. This was particularly true during the Russian interference, when the use of trolls and bots would have increased the illusion of consensus for human members of affected Facebook Groups.

People in filter bubbles can be manipulated. They share at least one core value with the members of their group. Having a shared value fosters trust in the other members and, by extension, the Group. When a Group member shares a news story, the other members will generally give it the benefit of the doubt. When Group members begin to embrace that story, the pressure mounts on other members to do the same. Now imagine a Facebook Group that includes a Russian troll and enough Russian-controlled bots to represent 1 to 2 percent of the membership. In that scenario, the trust of Group members leaves them particularly vulnerable to manipulation. A story like Pizzagate might get its start with the troll and bots, who would share it. Recipients assume the story has been vetted by members of a shared political framework, so they share it, too. In this way, disinformation and conspiracy theories can gain traction quickly. The Russians used this technique to interfere in the 2016 election. They also organized events. A particularly well-known example occurred in Houston, Texas, where Russian agents organized separate events for pro- and anti-Muslim Facebook Groups at the same mosque at the same time. The goal was to trigger a confrontation.

All of this is possible because users trust what they find on social media. They trust it because it appears to originate from friends and, thanks to filter bubbles, conforms to each user's preexisting beliefs. Each user has his or her own *Truman Show*, tailored to press emotional buttons, including those associated with fear and anger. While endless confirmation of their preexisting beliefs sounds appealing, it undermines democracy. The Russians exploited user trust and filter bubbles to sow discord, to reduce faith in democracy and government, and, ultimately, to favor one candidate over the other. The Russian interference succeeded beyond any reasonable expectation and continues to succeed because many key stakeholders in our government have been slow to acknowledge it, taking no meaningful steps to prevent a reoccurrence. Filter bubbles and preference bubbles undermine critical thinking. Worse still, the damage can persist even if the user abandons the platform that helped to foster them.

Listening to Renée describe the techniques employed by the Russians, I realized I was in the presence of a genuine superstar. At the time, I had only the barest outline of her life story, but the elements I knew—technology operations for the government, algorithmic trading, political campaigns, infiltration of antivaxx networks, harassment on Twitter, research on Russian interference in democracy—spoke to Renée's intelligence and commitment. Adding Renée to our team was transformational. Unlike Tristan and me, Renée could go beyond hypotheses. She had been researching this stuff for several years. She lived in the world of facts. Every member of Congress and staffer we met took to her immediately. What Renée got from us was a new platform to share what she knew. She had been working at Data for Democracy, identifying threats early and asking the tech companies to take the spread of computational propaganda more seriously. As so often happens with researchers and intelligence professionals, their brilliant insights did not always reach the right people at the right time. It was obviously too late to stop interference in the 2016 election, but perhaps our little trio could help prevent repeats in 2018 and 2020. That was the goal.

Our meeting concluded with a request from the House Intelligence Committee staffers: Could we help them learn about Facebook and Twitter? The committee was planning to have a hearing on the same day as the Senate Intelligence Committee, calling the same witnesses. But the staff needed help. Due to the sensitive nature of their work, none of the staff members used social media intensely. They needed to learn the inner workings of Facebook, Instagram, YouTube, or Twitter. Could we prepare briefing materials? Could we prepare questions the members could ask the witnesses? Renée and I jumped at the opportunity. We had seven weeks to create a curriculum and teach the staff Internet Platforms 101.

The opportunity went beyond the minority staff for the House Permanent Select Committee on Intelligence. Senator Warner's staff had asked for similar help, as did Senators Richard Blumenthal, Al Franken, Amy Klobuchar, and Cory Booker, all of whom served on the Senate Judiciary Committee, which would hold its hearing the day before the intelligence committees.

The process of creating briefing materials was iterative, with many revisions. Staffers posed questions several times a week, using our insights about how the platforms worked to understand intelligence gathered from their sources that they never shared with us. They asked for a briefing on algorithms and how they work in the context of Facebook, in particular. They understood that an algorithm is, as Wikipedia defines it, "an unambiguous specification of how to solve a class of problems," generally related to calculation, data processing, and automated reasoning. But with advances in AI, algorithms have become more complex as they adapt, or "learn," based on new data. In Facebook's relentless effort to eliminate any friction that might limit growth, they automate everything, relying on ever-evolving algorithms to operate a site with 2.4 billion active users and millions of advertisers. Algorithms find patterns shared by different users, based on their online behavior. This goes way beyond their common interests to include things like the time, location, and other context elements of web activity. If User A does a dozen things online prior to buying a new car—many of them unrelated to buying a car—the algorithm will look for other users who start down the same path and then offer them ads to buy a car. Given the complexity of Facebook, the AI requires many algorithms, the interaction of which can sometimes produce unexpected or undesirable outcomes. Even the smallest changes to one algorithm can trigger profound ripple effects through the rest of the system. A clear case where moving fast can break things in unpredictable ways.

While Facebook argues that its technology is "value neutral," the evidence suggests the opposite. Technology tends to reflect the values of the people who create it. Jaron Lanier, the technology futurist, views the role of algorithms as correlating data from individual users and between users. In an opinion piece in *The Guardian*, Lanier wrote, "The correlations are effectively theories about the nature of each person, and those theories are constantly measured and rated for how predictive they are. Like all well-managed theories, they improve through adaptive feedback." When it comes to the algorithms used by internet platforms, "improve" refers to the goals of the platform, not the user. Algorithms are used throughout the economy to automate decision making. They are authoritative, but that does not mean they are fair. When used to analyze mortgage applications, for example, algorithms that reflect the racial biases of their creators can and do harm innocent people. If the creators of algorithms are conscious of their biases and protect against them, algorithms can be fair. If, as is the case at Facebook, the creators insist that technology is by definition value neutral, then the risk of socially undesirable outcomes rises dramatically.

Given the importance of presidential elections to our democracy, the country had every right to insist that the CEOs of Facebook, Google, and Twitter testify. The CEOs of any other industry would have been there. But that did not happen with Facebook, Google, and Twitter. For reasons that underscore the partisan divide in Congress, the majority did not insist on testimony from the CEOs. They settled for the general counsel of each company. These are very well-educated, successful lawyers, but they are not engineers. They are good at talking, but their familiarity with the inner workings of their firm's products would have been limited. If the goal was to minimize the effectiveness of the hearings from the outset, they were the perfect witnesses. They were there to talk without saying anything, to avoid blunders. Knowing this, we provided a long list of ways that Facebook, Google, and Twitter might deflect questions at the hearing, along with comebacks to get at the key information.

In Silicon Valley, it is widely assumed that the government does not function well. The perception is that the only great people working in Washington are the ones who went there from Silicon Valley. Our experience could not have been more different. The staffers with whom we worked during this period were uniformly impressive. It's not just that they were smart, conscientious, and hard-working. Like users, policy makers had trusted Silicon Valley to regulate itself, so they had work to do to get up to speed for oversight. They knew what they did not know and were not afraid to admit it.

The night before the first hearing, Facebook disclosed that 126 million users had been exposed to Russian interference, as well as 20 million users on Instagram. Having denied any role in the Russian interference campaign for eight months, only to concede that an internal investigation had uncovered one hundred thousand dollars' worth of Russian advertising purchases in rubles, this revelation came as a bombshell. The user number represents more than one-third of the US population, but that grossly understates its impact. The Russians did not reach a random set of 126 million people on Facebook. Their efforts were highly targeted. On the one hand, they had targeted people likely to vote for Trump with motivating messages. On the other, they identified subpopulations of likely Democratic voters who might be discouraged from voting. The fact that four million people who voted for Obama in 2012 did not vote for Clinton in 2016 may reflect to some degree the effectiveness of the Russian interference. How many of those stayed away because of Russian disinformation about Clinton's email server, the Clinton Foundation, Pizzagate, and other issues? CNN reported that the Russians ran a number of Facebook Groups targeting people of color, including Blacktivist, which gained a substantial following in the months before the election. They ran another Group called United Muslims of America, with a similar approach to a different audience. On Twitter, the Russians ran accounts like "staywoke88," "BlackNewsOutlet," "Muslimericans," and "BLMSoldier," all designed, like Blacktivist and United Muslims of America, to create the illusion that genuine activists supported whatever positions the Russians promoted. In an election where only 137 million people voted, a campaign that targeted 126 million eligible voters almost certainly had an impact. How would Facebook spin that?

The hearings began with Senate Judiciary on Halloween. We knew that the staffers had reached out to many other people, and it was fun to hear some of our questions. The general counsels of Facebook, Google, and Twitter stuck to their scripts. Not surprisingly, Facebook's general counsel, Colin Stretch, faced the toughest questions, nearly all of which came from Democrats. Stretch held his own until relatively late in the hearing, when Senator John Kennedy, a Republican from Louisiana, surprised the whole world—and especially some of his Republican colleagues—by posing a series of questions to Stretch about whether Facebook had the ability to look at personal data for individual users. Stretch tried to deflect the question by answering that Facebook had policies against looking at personal data. Hidden behind an "aw shucks, I'm a country lawyer" demeanor, Kennedy has a brilliant mind, and he reframed his question until Stretch was forced to answer yes or no. Did Facebook have the ability to look at a user's personal data? Kennedy reminded Stretch he was under oath. "No," was Stretch's final answer. In my head, I heard the game-show sound for a wrong answer. *Baaamp.* To me, this was a big deal. Facebook is a computer system with oceans of data. Facebook engineers must have access to the data to do their jobs. In the company's early years, it was not uncommon for a Facebook recruiter to access a candidate's page during an interview. At some point, though, the company recognized that access to individual accounts had to be off-limits. They made clear to employees that inappropriate access would result in immediate termination. I do not know how effectively Facebook policed the rule—at Facebook, rules often are more about preventing liability than about changing behavior—but the statement that no one at Facebook could access individual data was incorrect. Next time Congress held hearings, the committee could point to Stretch's testimony as evidence of the need for testimony from executives higher up the organizational chart than the general counsel. It provided grounds for congressional committees to insist that CEOs testify in a future hearing, should the political winds change. It was a small victory.

In the end, Stretch and the other general counsels managed to dodge the toughest questions posed by the Senate Judiciary Committee, eliciting a rebuke from the ranking member, Senator Dianne Feinstein. When the Senate Intelligence hearing began the next morning, the dance continued. I could not help but be impressed by the way the general counsels sidestepped tough questioning with answers that sounded reasonable but were actually content-free.

The final hearing, House Intelligence, was surreal. The Republican members, under the leadership of committee chair Devin Nunes, were holding one kind of hearing, while the Democrats, led by ranking member Adam Schiff, held another. Our interest was in the Schiff-led hearing. You would not think there would be anything left for House Intelligence, following the two Senate hearings, but there was. House Intelligence showed examples of Facebook ads run by Russian-backed groups, which were printed in a large format for display during the hearing. The images of those ads remain indelible in my mind, the defining visual from the hearings.

Televised congressional hearings are typically long on theater and short on substance. That is especially true in an environment as polarized as Capitol Hill in 2017. From our perspective, the normal rules did not apply to these hearings. By making millions of Americans aware that internet platforms had been exploited by the Russians to interfere in our presidential election, these hearings had done something important. They were a first step toward congressional oversight of the internet platforms, and we had played a small role. A few hours after the House Intelligence hearing, I received an email that made me laugh. Sent by the lead investigator for Democrats on House Intelligence, it read, "In a workplace appropriate manner, I love you."

Other than Stretch's misstatement to Senator Kennedy, the general counsels succeeded in their mission. They did not add fuel to the fire, but the revelations about Facebook had a long hangover. There was an explosion of press interest in the three companies' role in the Russian interference. By televising the hearings, the congressional committees ensured at least one news cycle of coverage. In combination with Facebook's revelation about the 126 million users exposed to Russian interference, the hearings accomplished more than that. The possibility that Facebook, Google, and Twitter had played a role in undermining democracy became a topic of conversation. I noticed a big uptick in interest from mainstream media outlets. On the afternoon of November 1, while the House Intelligence hearing was unfolding, I made my first appearance on MSNBC, talking to Ali Velshi about these issues. It would be the first of many. People were beginning to talk about the role of Facebook in the election interference of 2016. The next day, Tristan and I were chatting on the phone, and he observed that we'd come a long way since April. In less than seven months, we had realized our goal of helping to trigger a serious conversation about the dark side of social media. It was a start, but nothing more. The platforms were still deflecting responsibility and that would not change without dramatically more public awareness and pressure.

7

The Facebook Way

*The problem isn't any particular technology, but the
use of technology to manipulate people, to concentrate
power in a way that is so nuts and creepy that it
becomes a threat to civilization.* —JARON LANIER

Thanks to the hearings, the press took a greater interest in the role of
internet platforms in the Russian interference. Every story added to
public awareness and gradually increased the pressure on policy makers
to do something. We met with many politicians, which helped us appreciate one of the rules of politics: if you want to bring about change
and don't have a huge lobbying budget, there is no substitute for pressure from voters. I paid five hundred dollars and attended a breakfast
for Senator John Kennedy of Louisiana the week after the hearings.
It was a breakfast with the senator, two staffers, nineteen lobbyists,
and me. Before the event started, the senator walked around the
room, greeting each person. The lobbyists were from companies like
Procter & Gamble, Alcoa, and Amazon. All they wanted to talk about
was the upcoming tax-cut bill. When the senator got to me, I said,

"I'm here in my capacity as a citizen to thank you for the amazing job you did at the Senate Judiciary hearing last week." Senator Kennedy did a double take. I no longer remember his exact words, but in a trademark drawl he told me, "Son, I appreciate that. I'm glad you're here. I want to meet with you again." Unfortunately, that has not happened yet.

During the second week of November, I participated in a conference in Washington on antitrust regulation, sponsored by the Open Markets Institute think tank. The keynote speaker was Senator Al Franken, who made a full-throated argument for traditional approaches to regulating monopolies. Columbia Law's Tim Wu and Open Markets' Lina Khan talked about antitrust in the context of internet platforms. They argued that Amazon, Google, and Facebook all have monopoly power that would not have been permissible for most of the twentieth century, and they use it to block competitors and disadvantage users. My own remarks framed Tristan's hypotheses about public health as an argument for antitrust regulation of internet giants. The internet still enjoyed exemptions from regulation that were artifacts of the industry's early days. Over the course of twenty-one years, regulatory safe harbors had enabled the market leaders not only to prosper but also to do things no other industry could get away with, including noncooperation with regulators like the Federal Trade Commission, a careless disregard for consumer data privacy, and an exemption from Federal Communications Commission rules for election-related advertising.

Until 1981, the United States operated with a philosophy that monopoly was bad for consumers and for the economy. Monopolies can charge higher prices to consumers than competitive markets, while also slowing the rate of innovation and new company formation. The rise of Standard Oil and other trusts around the turn of the twentieth century created the impetus for the Sherman Antitrust Act, the Clayton Act, and the Federal Trade Commission Act, which ushered in a long period when both political parties supported efforts to prevent anticompetitive concentration of economic power. A counter philosophy surfaced after the Second World War, which postulated that markets were always best at allocating resources. The "Chicago School" antitrust philosophy emerged as part of this market-driven, neoliberal worldview, arguing that concentration of economic power was not a problem, so long as it did not translate into higher prices for consumers. The Chicago School became official policy with the Reagan administration and has prevailed ever since. Perhaps it is a coincidence, but, as I've mentioned, the years since 1981 have seen a massive decline in new company formation (which peaked in 1977), as well as income inequality not seen since the era of Standard Oil.

Three internet platforms—Amazon, Google, and Facebook—have benefited enormously from the Chicago School's antitrust philosophy. The products of Google and Facebook are free to consumers, and Amazon has transformed the economics of distribution while keeping consumer prices low, which has allowed all three to argue successfully for freedom to dominate, as well as to consolidate. The case against Amazon is probably strongest, and it provides a framework for understanding the larger issues.

For Amazon.com, freedom from antitrust scrutiny has allowed the company to integrate vertically, as well as horizontally. From its original base in retail for nonperishable goods, Amazon has expanded horizontally into perishables, with Whole Foods, and into cloud services, with Amazon Web Services. Amazon's vertical integration has included Marketplace, which incorporates third-party sellers; Basics, where Amazon private-labels bestselling commodity products; and hardware, such as Alexa voice-controlled devices and the Fire home video server. In a traditional antitrust regime, Amazon's vertical-integration strategy would not be allowed. The use of proprietary consumer data to identify, develop, and sell products in direct competition with bestsellers on the site represents an abuse of power that would have appalled regulators prior to 1981. Amazon's ever-expanding distribution business might have run afoul of the same concerns. The horizontal integration into perishables like food would have been problematic due to cross subsidies. Amazon can use its cloud services business to monitor the growth of potential competitors, though there is little evidence that Amazon has acted on this intelligence the way it has leveraged data about bestselling products in its marketplace.

Google's business strategy is a perfect example of how the Chicago School differs from the traditional approach to antitrust. The company began with index search, arguably the most important user activity on the internet. Google had a brilliant insight that it could privatize a large subset of the open internet by offering convenient, easy-to-use, free alternatives to what the web's open source community had created. Google leveraged its dominant market position in search to build giant businesses in email, photos, maps, videos, productivity applications, and a variety of other apps. In most cases, Google was able to transfer the benefits of monopoly power from an existing business to a nascent one. The European Union, which still employs a traditional view of economic power, won a $2.7 billion judgment against Google in 2017 for leveraging its search and AdWords data to wipe out European competitors for its brand-new price-comparison application. The EU case was well argued and had the benefit of obvious harm, in that most of Google's competition had disappeared in short order. Shareholders shrugged off the judgment, which Google has appealed. (In August 2018, the EU fined Google $5 billion for a different antitrust violation, this time related to the Android operating system.)

The Chicago School antitrust model benefited Google and Facebook in another way: it enabled them to create markets in which they could also be a participant. Traditional antitrust rules offered companies a choice: they could create a market or be a participant but not both. The theory was that if the owner of a marketplace also participated in the market, the other competitors would be at a prohibitive disadvantage. Google's purchase of DoubleClick enabled precisely this situation in the online advertising business, enabling Google to favor its own properties at the expense of third parties. Google did something similar when it acquired YouTube, changing the algorithm in ways that, among other things, gave its own content preferred distribution.

Google and Facebook operate a form of what economists call "two-sided markets," which Wikipedia defines as "economic platforms having two distinct user groups that provide each other with network benefits." The original two-sided markets included things like credit cards where the issuer sits between vendor and customer in a single transaction. For platforms, the two sides are not part of the same transaction. Users are the source of data, as well as the product, but they do not participate in a transaction or in the economics. The advertisers are the customer and provide the market's revenue. What makes the platforms comparable to traditional two-sided markets is that both sides depend on the success (or scale) of the market. At the scale of Facebook and Google, the two-sided market confers advantages that cannot be overcome by competitors, providing monopoly power.

Thanks to its search engine, cloud services, and venture capital operation, Google has an exceptionally good view of emerging products. Google has never been shy about using its market power to limit the upside of new companies, acquiring the best and snuffing out the rest. American regulators do not see any problem with this behavior. The European regulators have been trying to rein in Google on their own.

Facebook has enjoyed similar advantages to Google from the Chicago School antitrust model. Facebook imitated Google's privatization of the open web, complementing its social network with a photos app (Instagram), text messaging (WhatsApp and Messenger), and virtual reality (Oculus). Having cornered the audience for content, Facebook has undercut the economics of journalism by seducing publishers to work with it on new products, like Instant Articles, and then changing the terms to the disadvantage of publishers. In 2013, Facebook acquired

Onavo, an Israeli company that makes a virtual private network (VPN) application. VPNs are a tool for protecting privacy on public networks, but Facebook has given Onavo a twist straight out of Orwell. Onavo enables Facebook to track everything the user does when using the VPN. Onavo also enables Facebook to surveil other applications. Neither of these activities would be considered appropriate for a VPN under normal circumstances. It is roughly analogous to a security service that protects your home from other thieves but steals your valuables while they are doing it. The whole point of a VPN is to prevent snooping. Enough users use Onavo to give Facebook huge amounts of data about users and competitors. In August 2018, Apple announced that Onavo violated its privacy standards, so Facebook withdrew it from the App Store.

One of the competitors Facebook has reportedly tracked with Onavo is Snapchat. There is bad blood between the two companies that began after Snapchat rejected an acquisition offer from Facebook in 2013. Facebook started copying Snapchat's key features in Instagram, undermining Snapchat's competitive position. While Snapchat managed to go public and continues to operate as an independent company, the pressure from Facebook continues unchecked and has taken a toll. Under a traditional antitrust regime, Snapchat would almost certainly have a case against Facebook for anticompetitive behavior.

Freedom from antitrust scrutiny has enabled the internet giants to dominate their markets to a degree unseen since the heyday of IBM's dominance in mainframe computers. In reality, today's internet platforms are far more influential than IBM in its prime. With 2.4 billion monthly users on its core platform, Facebook directly influences nearly one-third of the world's population. Other Facebook platforms also have huge monthly user bases: 1.5 billion use WhatsApp, 1.3 billion use Messenger, and 1 billion use Instagram. While there is overlap, particularly between Messenger and Facebook, both WhatsApp and Instagram have large numbers of users who are not on Facebook. The company announced that it has 2.8 billion users across its core platforms, or nearly 40 percent of the world's population. On its best day, IBM's monopoly was limited to governments and the largest corporations. Thanks to brain hacking and the filter bubbles that result from it, Facebook's influence over consumers may be greater than any single business before it.

Persuading 40 percent of the world's population to use your products is an extraordinary accomplishment. In many circumstances, it would be entirely laudable. For example, Coca-Cola serves 1.9 billion beverages per day across two hundred countries. But Coca-Cola does not influence elections or enable hate speech that leads to violence. As a giant communications network, Facebook has far more influence than Coca-Cola, and unlike Coca-Cola, Facebook has monopoly power. With such influence and monopoly power should come great responsibility. Facebook owes a duty to its users—and the whole world—to optimize itself for the public good, not just for profits. If Facebook cannot do that—and the evidence at this point is not promising—then government intervention to reduce its market power and introduce competition will be required.

Zuck's and Sheryl's failure to take action to address obvious flaws in the product and to protect their brand is at least suggestive of their monopoly power. They may not have been concerned about brand damage because they knew that users had no alternative.

There is a second possible explanation for why Facebook might have ignored early warnings and later criticism. From his time at Harvard, Mark Zuckerberg showed a persistent indifference to authority, rules, and the users of his products. He hacked servers at Harvard, took university property to create his first products, exploited the trust of the Winklevoss brothers, and then shared his view of users in an instant messaging exchange with a college friend just after the launch of TheFacebook. As quoted in *Business Insider*:

Zuck: Yeah so if you ever need info about anyone at Harvard

Zuck: Just ask.

Zuck: I have over 4,000 emails, pictures, addresses, SNS

[Redacted Friend's Name]: What? How'd you manage that one?

Zuck: People just submitted it.

Zuck: I don't know why.

Zuck: They "trust me"

Zuck: Dumb fucks.

As far as I can tell, Zuck has always believed that users value privacy more than they should. As a result, he has generally chosen to force them to be more open and then dealt with the fallout when it came. For the most part, the bet against privacy paid off for Facebook. Negative user feedback forced Facebook to withdraw Beacon, but the company's relentless efforts overwhelmed resistance far more often than not. Users either did not know or did not care about the loss of privacy, enabling Facebook to join the list of most valuable companies on earth.

Facebook's motto, "Move fast and break things," reflects the company's strengths and weaknesses. Facebook constantly experiments, tinkers, and pushes envelopes in the pursuit of growth. Many experiments fail or work imperfectly, necessitating an apology and another experiment aimed at doing better. In my experience, there have been few, if any, companies that have executed a growth plan—moving fast, if you will—as effectively as Facebook. When moving fast leads to breaking things, and to mistakes, Facebook has been brilliant in its ability to recover from them. Seldom has Facebook allowed a mistake or problem to slow it down. Most of the time, promises to do better have been enough to get past a problem.

To be clear, I believe that taking risks is a positive thing in business when accompanied by good judgment. Where Facebook failed was in not recognizing that tactics need to change as a company's influence grows. Experiments that are acceptable at small scale can be problematic at a larger one. When a company reaches global scale, as Facebook has done, it needs to approach experimentation with extreme care. It must prioritize users and the public interest. It must anticipate and prepare for side effects.

One of the things that distinguished Zuck from the beginning was his vision that Facebook could connect the entire world. When I knew him, Zuck had his eyes set on reaching one billion users. At this writing, the company has 2.4 billion monthly users. Revenues in 2018 approached fifty-six billion dollars. To reach those numbers in fifteen years from a standing start required more than brilliant execution. There were costs, borne by others. Facebook eliminated all forms of friction that might have slowed it down, an activity that Zuck and his team have transformed into a fine art. Regulation and criticism? Facebook makes them disappear with the magic words "We apologize. We'll do better!" Too often, those words have not been matched to action, as that kind of action would have slowed down the company. And there is almost no way for regulators or critics to verify Facebook's compliance. Until recently, the company resisted all attempts at transparency with respect to its algorithms, platforms, and business model. The company's 2018 labeling requirement for political ads was a step in the direction of transparency. Unfortunately, Facebook's implementation has fallen far short of what is necessary.

Even at huge scale, Facebook's business is relatively straightforward. In comparison to a similarly sized business, such as the Walt Disney

Company, Facebook is operationally far less complex. The core platform consists of a product and a monetization scheme. The acquired products—Instagram, WhatsApp, and Oculus—operate with a fair amount of autonomy, but their business models add little complexity. The relative simplicity of the business enables Facebook to centralize its decision making. There is a core team of roughly ten people who manage the company, but two people—Zuck and Sheryl Sandberg—are the final arbiters of everything. They have surrounded themselves with a team of brilliant operators who executed the strategy of maximum growth almost flawlessly through the end of 2017.

Thanks to Facebook's extraordinary success, Zuck's brand combines elements of rock star and cult leader. He is deeply committed to products and not much interested in the rest of the business, which he leaves to Sheryl. According to multiple reports, Zuck is known for micromanaging products and for being decisive. He is the undisputed boss. Zuck's subordinates study him and have evolved techniques for influencing him. Sheryl Sandberg is brilliant, ambitious, and supremely well organized. When Sheryl speaks, she chooses her words very carefully. In an interview, for example, she has mastered the ability to appear completely genuine and heartfelt, while being totally nonresponsive. When Sheryl talks, friction disappears. She manages every detail of her life, paying particular attention to her image. Until mid-2018, Sheryl had a consigliere, Elliot Schrage, whose title was vice president of global communications, marketing, and public policy, but whose real job appeared to be protecting Sheryl's flank, something he had done since her time at Google.

If you wanted to draw a Facebook organizational chart to scale, it would look like a large loaf of bread with a giant antenna pointing straight up. Zuck and Sheryl are at the top of the antenna, supported by Schrage and product boss Chris Cox until they departed, the company's chief financial officer David Wehner, and a handful of others. Everyone else is down in the loaf of bread. It is the most centralized decision-making structure I have ever encountered in a large company, and it is possible only because the business itself is not complicated. Early in Sheryl's tenure at Facebook, something happened that revealed her management philosophy to me. The context was a failure of judgment that in most companies would have resulted in the termination of the person who made the decision and changes in policy. I called Sheryl to ask her how she planned to handle it, and she said, "We are a team at Facebook. When we succeed, we do so as a team. When we fail, it is a team failure." When I pushed back, Sheryl did the same. "Are you saying you want me to fire the entire team?" In retrospect, that might have been for the best.

The management philosophy that Sheryl described has huge benefits when everything is going well because it keeps everyone focused on their metrics rather than on self-promotion. In Facebook's case, everything went perfectly from shortly after the IPO in 2012 until the end of 2017. Imagine a finely tuned race car zooming down a straightaway with no pebbles on the road. That was Facebook. Inevitably, something will go wrong. That is the real test. In theory, the team philosophy might create a safe space for disagreement and self-examination, but that is not what happened at Facebook. When there is no individual credit for members of the team, much of the credit for every success goes to the people at the top. Given Zuck's status as the founder, the team at Facebook rarely, if ever, challenged him on the way up and did not do so when bad times arrived. This reveals the downside of Sheryl's management philosophy: no credit/no blame can eliminate accountability for mistakes. When faced with a setback, the team may circle the wagons and deflect criticism rather than do any soul-searching. That appears to be what happened to Facebook when confronted with evidence that its platform had been exploited by the Russians.

Facebook has made many mistakes in its history, but the Russian interference was the first that could not be easily dismissed. It created friction unlike anything Facebook had encountered in its first thirteen years. The company had no experience—no muscles—for dealing with that kind of friction. They rolled out their standard response—deny, delay, deflect, dissemble—expecting the friction to go away. It always had in the past. But this time, the friction remained. Perhaps it might evaporate at some point, but not so quickly as in the past. With no one to push back on Zuck and Sheryl, Facebook stuck with its playbook, doing the same thing over and over, expecting a different result. Unused to negative feedback, Zuck and Sheryl retreated into a bunker. They reappeared only when there was no alternative.

Tristan, Renée, and I watched in wonder as Facebook executed its strategy of deny, delay, deflect, dissemble. How could Zuck and Sheryl not see where this was heading? We were a tiny team with few resources, but we were not alone any longer. Election interference was an issue whose time had come. A lot of smart people were looking at the problem from different angles. We just happened to be in the right place when the story hit, with growing demand for our perspective from policy makers and journalists. Fortunately, two recruits joined us in November, giving us new skills and new energy. Lynn Fox, who had been a senior communications executive at Apple, Palm, and Google, brought expertise in media that would soon transform our effort. At mid-month, *The New York Times* wrote a profile about Renée. A week later, a former Facebook privacy manager named Sandy Parakilas wrote an op-ed in the *Times* entitled, "We Can't Trust Facebook to Regulate Itself." Chris Kelly, the original chief privacy officer at Facebook who had introduced me to Zuck in 2006, knew Sandy and made that introduction. Sandy was aware of what Tristan, Renée, and I were doing and asked if he could join forces with us. Sandy would prove to be an exceptionally fortuitous addition to our team. When we met, Sandy was a former Facebook employee; four months later, events would transform him into a whistle-blower.

On December 11, 2017, *The Verge* reported that Chamath Palihapitiya, Facebook's former vice president of growth, had given a speech at

Stanford the month before in which he had expressed regrets about the negative consequences of Facebook's success. "I think we have created tools that are ripping apart the social fabric of how society works," he told the students at the Graduate School of Business. Palihapitiya's remarks echoed those of Sean Parker, the first president of Facebook, who in November had expressed regret about the "social-validation feedback loop" inside the social network, which gives users "a little dopamine hit every once in a while, because someone liked or commented on a photo or a post or whatever." Facebook had ignored Parker, but apparently they jumped on Palihapitiya. Within seventy-two hours of *The Verge*'s initial report, Palihapitiya publicly reversed course. "My comments were meant to start an important conversation, not to criticize one company—particularly one I love. I think it's time for society to discuss how we use the tools offered by social media, what we should expect of them and, most importantly, how we empower younger generations to use them responsibly. I'm confident that Facebook and the broader social media category will succeed as they navigate this uncharted territory." He subsequently appeared on Christiane Amanpour's show on CNN International and made it clear he thought Mark Zuckerberg was the smartest person he had ever met and suggested that Zuck was uniquely qualified to figure it out and save us all.

I don't really know Chamath. I have had only one substantive conversation with him, for ninety minutes in 2007, when Zuck asked me to help recruit Chamath to Facebook. At the time, Chamath was working at the venture firm Mayfield Fund, whose office was one flight up from mine at Elevation. Chamath was born in Sri Lanka and emigrated with his family to Canada. He overcame economic challenges, got a first-class education, and made his way to Silicon Valley. Brilliant, hard-working, exceptionally ambitious, and confident that his actions would always be right, Chamath exudes the vibe of classic Silicon Valley bro. He is also a very successful poker player, having once placed 101st out of 6,865 participants in the World Series of Poker's Main

Event. In short, Chamath Palihapitiya is no shrinking violet. He is not the sort of person to back down because someone yells at him. And yet, he went from being an articulate critic of Facebook to a willing purveyor of the company's PR lines almost overnight. It was enough to make one suspicious.

Why was Chamath's criticism more problematic for Facebook than that of Sean Parker or any of the earlier critics? There was one obvious difference. Before Chamath left Facebook in 2011, he had recruited many of the leaders of the Growth team. In Facebook parlance, Growth is about all the features that enable the company to increase user count and time on site and sell ads so successfully. (In Tristan's framing, Growth is the group responsible for brain hacking.) If Chamath had continued to question Facebook's mission, it is quite possible that the people he hired at the company, and those who knew him, might begin to question their leaders' and company's choices. The result might be a Susan Fowler Moment, named for the Uber engineer whose blog post about that company's toxic culture led to an employee revolt and, ultimately, the departure of the executive team. What made Fowler so important was that Uber's management team, board of directors, and investors had done nothing for years to change the toxic culture, despite a steady flow of bad news about it. Fowler framed the problem in a way that no one could deny, causing the employees to demand change. That is precisely what Palihapitiya did in his remarks at Stanford. It's easy to imagine that Facebook might do whatever it took to prevent a Susan Fowler Moment.

Chamath's reversal triggered an insight. The window for Facebook to make a graceful exit from its predicament would not stay open forever. The opportunity to follow the example of Johnson & Johnson after the Tylenol tampering crisis of the 1980s would last only as long as Facebook could credibly plead ignorance of its misdeeds. Chamath had presented Facebook with a teachable moment. They could have said, "Now we get it! We screwed up! We will do everything possible to fix the problems and restore trust." By failing to exploit Chamath's regrets as a teachable moment, Facebook signaled a commitment to avoiding responsibility for the Russian election interference and all the other problems that had surfaced. This was bad news. I had been giving Facebook the benefit of the doubt since October 2016, assuming that the company had been a victim. For six months after my original email to Zuck and Sheryl, I had assumed that my delivery was flawed or that I had been the wrong messenger. When Tristan and I started speaking out, I hoped that Facebook employees and alumni would join the cause and that people like Sean Parker and Chamath Palihapitiya might succeed in convincing Zuck and Sheryl to change their approach. That did not happen.

8

Facebook Digs in Its Heels

Success in creating AI would be the biggest event in
human history. Unfortunately, it might also be the last,
unless we learn how to avoid the risks. —STEPHEN HAWKING

When a company grows from nothing to 2.4 billion active users and fifty-six billion dollars in revenues in only fifteen years, you can be sure of three things: First, the original idea was brilliant. Second, execution of the business plan had to be nearly flawless. And third, at some point along the way, the people who manage the company will lose perspective. If everything your company touches turns into gold for years on end, your executives will start to believe the good things people say about them. They will view their mission as exalted. They will reject criticism. They will ask, "If the critics are so smart, why aren't they as successful and rich as we are?"

Companies far less successful than Facebook have fallen victim to such overconfidence. The culture of Silicon Valley, which celebrates the brash and the bold, breeds overconfidence and then lets nature take its course. The corpses of overconfident companies litter the landscape of the tech industry. Companies like Digital Equipment, Compaq, Netscape, Sun Microsystems, and MySpace were hot growth stories in their prime. Then there are the survivors who had lost prestige as their growth slowed down, companies like Intel, EMC, Dell, and Yahoo. The executives of these companies confidently predicted strong growth right up to the moment when there was none and the stock—and their dreams of endless growth—came tumbling down. Then there are companies like Oracle and IBM, at one time undisputed leaders but whose giant market capitalization masks a dramatic loss in influence.

It took me a very long time to accept that Zuck and Sheryl had fallen victim to overconfidence. I did not pick up the signal when I first reached out to them in October 2016. It was not even clear to me in February 2017, when I gave up on trying to convince Dan Rose to investigate my concerns. The evidence started to pile up in the month prior to the hearings on October 31 and November 1, 2017, but I still wanted to believe that Zuck and Sheryl would eventually change their approach. The clincher was the episode with Chamath. Before Chamath, Facebook could have said, "We didn't cause the Russian interference, but it happened to our users and we will do everything in our power to protect them." Facebook could have followed the crisis management playbook, cooperated fully and enthusiastically with investigators and reached out to users who were touched by the Russian interference with an explanation and evidence. A lot of time had passed, but I am pretty sure that everyone who mattered—users, advertisers, the government, Facebook employees—would have reacted favorably. There might have been a hit to earnings and the stock price in the short term, but before long Facebook would have enjoyed the benefits of greater trust from its constituents, which would have taken the stock to new highs.

Instead, Facebook finished 2017 as it had begun it, by not giving an inch, thus violating a central precept of crisis management: embracing criticism. Instead, Facebook defied its critics without even acknowledging their existence. The company's message to the world—"nothing to see here, move along"—was so completely out of step with what we already knew that I was taken aback. What were they thinking?

The idealist in me still hoped there would be a way to persuade Zuck and Sheryl to look at the situation differently, to recognize that the 2018 US midterm elections were fast approaching and only Facebook had the ability to protect that election from mischief similar to 2016.

From the start, we understood that all the best outcomes required cooperation from Facebook. Election interference and threats to public health were the result of Facebook's design choices, made in service of a brilliantly successful advertising business model. Regulation could change Facebook's incentives and behavior, but that would almost certainly take years to implement, even in a best-case scenario, and that scenario was not really an option. Our only choices were to persuade Facebook to harm its business for the good of the country and the world—which seemed about as likely as my winning the hundred-meter dash in the Olympics—or just hope for the best.

Tristan remained optimistic that we might find allies among Facebook employees or alumni who could influence Zuck and Sheryl. No matter the odds, it was worth a try. No current employees were speaking out yet, but Tristan did not give up on them. He secured a series of meetings with influential Facebook executives who expressed interest in humane design, but so far that interest has not translated into changes in policy, presumably because only Zuck and perhaps Sheryl have the power to do that.

We had achieved our original goal of helping to trigger a conversation about the dark side of social media, but that had not yet produced any substantive change by the internet platforms. Our allies—especially those in Washington and in the media—encouraged us to keep pushing. They argued that Facebook would respond only to pressure, and they thought we could play a role in bringing pressure to bear. Facebook's strategy was to outlast our outrage, to concede nothing, and to hope the press and Congress would move on to other topics. Given the distraction of the Trump administration, that seemed like a good bet. Facebook had built up so much goodwill with users and politicians that it was understandable that they would expect the pressure to fade and disappear. Facebook had been overstepping boundaries and apologizing since Zuck's days at Harvard, and apologizing had always been enough to make past problems go away. Facebook was confident, but it was not invulnerable. The company's decision-making process, dominated by Zuck and Sheryl, appeared to be suboptimal for a crisis, which is where the election-interference story appeared to be going.

I had some perspective on an analogous situation. In 1994, Bill Gates asked me to be an early reader and sounding board for his first book, *The Road Ahead*. In those days, I paid exceptionally close attention to all things Microsoft. Not long thereafter, the US Justice Department initiated an antitrust case against Microsoft, citing unfair bundling of the Internet Explorer browser with Windows.

In those days, Microsoft was a global powerhouse equivalent to Google today, dominant and unassailable by competitors. A relatively small team in Redmond, Washington, made all the decisions. As an early adopter of electronic mail, Microsoft managed its global operations with the shortest lag times imaginable. Email enabled Microsoft employees in the remotest parts of Australia, South America, Africa, or Asia to escalate problems through the chain of command to the proper decision maker in Redmond in a matter of hours. It is hard to overstate the significance of the breakthrough represented by Microsoft's email system and the competitive advantage it provided. Until the antitrust case. The first thing Microsoft's antitrust lawyers did was to force a change in email practices to reduce the possibility of legal jeopardy. Overnight, a lawyer's decree converted internal communications at

Microsoft from an asset to a liability. The lawyers effectively decapitated the company. Microsoft probably would have messed up the internet opportunity in any case, but thanks to the antitrust case, they missed it by a mile.

If anything, decision making at Facebook was even more centralized than at Microsoft. Employees at Microsoft revered Bill Gates, but Bill encouraged debate. He was famous for saying, "That is the stupidest thing I have ever heard!" . . . but it was an invitation. He expected you to defend your position. If you did so well, Bill would come around. It's possible that something like that could happen at Facebook, but that did not appear to be the norm. At Facebook, Zuck is on a pedestal. I don't think many people debate or contradict him.

Back in early October 2017, Barry Lynn of the Open Markets Institute had convinced me to write a long-form essay describing the dark side of social media, our journey to date, and our best policy ideas. He then persuaded the editors of *Washington Monthly*, a venerable magazine for progressive policy advocates, to commission a 6,000-word essay. My goal was to crystallize the issues and our policy recommendations for policy makers inside the Beltway, an intellectual one-stop-shop that would frame the debate and create a platform for the next phase of our work. I submitted the first draft in late October.

I would get a master class in the Socratic method from the piece's editors, Paul Glastris and Gilad Edelman, who provided a series of brilliant questions, the answers to which filled in many holes. We showed the policy prescriptions to a number of others on Capitol Hill to get their feedback. The editing and reviewing went on through the month of November until one day Gilad told me we were done. The essay would be the cover story of the January 2018 issue, slated for publication on January 8. It went beyond election interference to talk about the threats from internet platforms to public health, privacy, and the economy.

PUBLIC HEALTH HAD BEEN Tristan's and my original focus, but Washington's interest in election integrity had dominated our attention since July. In late 2017, we resumed our public health effort, starting with the impact of technology products on children. In the mindless pursuit of growth, internet platforms had built a range of products for kids. It is hard to know whether the platforms were ignorant of children's vulnerability or drawn to it, but the kids' products they created appeared to cause developmental and psychological problems. On this issue, we found a terrific partner in Common Sense Media, the largest nonprofit focused on children and media in the United States. We had first talked to Common Sense in the summer of 2017. In addition to providing parents with reviews of television, films, and video games, Common Sense embraced the challenge of protecting children from age-inappropriate content on the web. This led to frequent and sometimes fierce battles with Facebook, Google, YouTube, Instagram, and Snapchat. I first met Common Sense founder Jim Steyer in tenth grade, and we have been friends ever since. From our first conversation about Tristan's work, Jim saw the opportunity for collaboration. Common Sense understood the threat posed by addiction to smart screens and was creating a series of public service announcements featuring comedian Will Ferrell to advocate for device-free dinners. The organization's biggest challenge was that its staff, well endowed with childhood development, policy, and lobbying skills, did not have enough credibility in the technology community. Without brand-name technologists in their team, Common Sense Media had less influence inside the industry than it wanted. In December 2017, after several months of conversations, Jim offered us a compelling package. Common Sense would allow us to use their offices for meetings and to leverage their considerable legislative prowess in Washington, DC, and the California statehouse in Sacramento. In exchange, Tristan would join Common Sense as a senior fellow and I as an advisor, bringing our technology experience and relationships to bear on their behalf. We planned to announce Tristan's new affiliation in Washington at a one-day conference on February 7. Common Sense's connections with policy makers and the media would draw attention to the event and increase the pressure on Facebook.

At the same time, Tristan initiated another effort. He wanted to bring technologists and other concerned parties together to create, enlarge, and strengthen the opposition to the business practices of internet giants. Having already built Time Well Spent into a vibrant community of people focused on taking control of their digital lives, Tristan understood that our advocacy needed a different kind of organization. He decided to create it: the Center for Humane Technology.

Tristan and I shared a view that human-driven technology could be the Next Big Thing in Silicon Valley. Tech products should not be dangerous. They should not misinform or dumb us down. The objective of new technology should be to empower users, enabling them to improve their lives. The current model cannot continue, but that does not mean that the technology industry has to suffer a decline. Like renewable energy—where solar and wind power have succeeded at utility scale—human-driven technology could replace an outdated approach with a new one, converting a man-made problem into a huge business opportunity. We want technology to make the world a better place. Human-driven technology is the way to do it.

What do I mean by human-driven technology? I want to see a return to technology that leverages the human intellect, consistent with Steve Jobs's "bicycle for the mind" metaphor. Human-driven products do not prey on human weakness. They compensate for weakness in users and leverage strengths. This means taking steps to prevent addiction and, when those fail, to mitigate the downsides. The design of devices should deliver utility without dependence. The design of applications and platforms should respect the user, limiting the effects of existing filter bubbles and preventing new ones from forming. Done right, every internet platform would be a new bicycle for the mind. In data privacy, a really useful idea would be a universal authentication system, an alternative to Facebook Connect or OpenID Connect that protects users. Facebook Connect is convenient for signing on to every site, but users need a way to do that without surrendering their privacy. The ideal would be to follow the model Apple set for facial recognition, where the data always remains on the smartphone, in the possession of the user. What I have in mind is an independent company that represents the interest of users at login, providing the minimum information required for each transaction.

With each new generation of technology, entrepreneurs and engineers have an opportunity to profit from designing products that serve rather than exploit the needs of their users. Virtual reality, artificial intelligence, self-driving cars, and the Internet of Things (IoT)—smart speakers and web-enabled televisions, automobiles, and appliances— all present opportunities to create bicycles for the mind. Unfortunately, I see no evidence yet that the designers in those categories are thinking that way. The term you hear instead is "Big Data," which is code for extracting value rather than creating it. At the end of the day, the best way to persuade Facebook and Google to adopt human-driven technology is to foster competition and demonstrate value in the marketplace. Giving consumers different design choices will take years, so we should start as soon as possible.

On January 1, 2018, Zuck published a post announcing his goal for the year ahead. It had become a tradition for Facebook's CEO to begin each year by giving himself a challenge. One year he learned to speak Mandarin. Another year he only ate meat he had killed himself. I have no idea why Zuck makes these challenges public. Zuck's goal for 2018 was to fix Facebook. He offered a nine-point plan. Wait. What? Fix Facebook? Where did that come from? No one at the company had previously admitted to problems that might require fixing. Suddenly

Zuck acknowledged concerns about fake news and the possibility that too much Facebook (or other social media) might lead to unhappiness. He offered a classic Zuck fix: more Facebook! Zuck's plan for addressing the problems created by Facebook was for users to do more of the things that created the problems in the first place. People who had not realized there was anything broken at Facebook were caught off guard by Zuck's post. What did it mean?

In response to Zuck's post, *Washington Monthly* published my cover story online a few days early, on January 5. Even though I had completed the essay more than a month earlier, it read like a rebuttal to Zuck's New Year's resolution. Where Zuck's post had framed Facebook's flaws obliquely, my essay was direct and specific. Where his remedy was more Facebook, my essay recommended ten remedies, aimed at user privacy, ownership of data, terms of service, and election interference. The press picked up on it, and the next thing I knew, an essay designed for a handful of policy makers in Washington began spreading outside the Beltway. We had managed to find a pretty good audience for our message in 2017, but we had made no progress with the only audience that mattered: the top people at Facebook. The company's PR people had begun to say unflattering things in private but were ignoring us in public. And it was working for them. Then came *Washington Monthly*. The lucky timing of the essay would lead Facebook to engage directly.

On Sunday, January 7, I flew to New York City with my wife, Ann, for a six-week stay. The next morning, I received an email from Jamie Drummond, Bono's partner on the ONE campaign, asking if he could introduce me to someone who represented George Soros, the billionaire investor who had dedicated his fortune to the promotion of democracy around the world. In my years in the investment business, there were a handful of people whose brilliance dazzled me, and George Soros was one. In an email, Soros's colleague Michael Vachon explained that Soros had read my *Washington Monthly* essay and liked it so much he planned to use it as the frame for a speech at the World Economic Forum in Davos on January 25. Would I be willing to meet with Mr. Soros to help him write the speech? I certainly would. We agreed to meet at the end of the week.

That same day, I called my friend Chris Kelly, the former chief privacy officer of Facebook, who had originally introduced me to Zuck. I wanted Chris's take on everything that we had learned. Chris shared my view that the only people who could fix Facebook quickly were Zuck and Sheryl. There was no way to protect elections or innocent people from harm without some cooperation from Facebook. Inside the company, Zuck and Sheryl had the power. They had the moral authority to change direction.

Unfortunately, Zuck and Sheryl refused to engage with critics. There are two reasons why crisis management experts advise clients to reach out to critics: you learn the dimensions of the problem, and by cooperating with critics, you take the first step on the path to restoring trust. Facebook had ignored all critics until Chamath Palihapitiya spoke at Stanford. Facebook's successful effort to get Chamath to recant his regrets and the resolution's focus on "more Facebook" as the solution to whatever ailed users suggested Facebook did not plan to concede anything.

The day took a turn into the twilight zone when a friend shared a tweet posted by long-time Facebook executive Andrew "Boz" Bosworth. It read, "I've worked at Facebook for 12 years and I have to ask: who the fuck is Roger McNamee?" It's a question I have often asked myself in other contexts, but in this case it pointed to an inescapable conclusion: the *Washington Monthly* essay had found its way into Facebook headquarters. He wasn't Zuck or Sheryl, but Boz was a member of the inner circle. We looked at the bright side. Getting on Boz's radar seemed like progress. It certainly caught the eye of many journalists, which helped our cause.

One thing had not changed from when I first reached out to Zuck and Sheryl in 2016: Facebook was not open to criticism, much less taking it to heart. Ignore the messenger was their first instinct; if that failed, presumably they would bring in the heavy artillery.

Later that day, Tim Berners-Lee tweeted the *Washington Monthly* essay to his followers. Berners-Lee is one of my heroes. His endorsement of the essay meant the world to me. Suddenly, the essay was everywhere, and requests came in from all sorts of media: CNBC, *Tucker Carlson* on Fox, *CBS Morning News*, *NBC Nightly News*, the *Today* show, MSNBC, *Frontline*, CNN, *60 Minutes*, *Bloomberg Technology*, BBC Radio, and Bloomberg Radio. Tristan, Sandy Parakilas, and I did a huge amount of TV and radio between Monday and Thursday of that week. Network television enabled us to share our message with millions of people. But the highlight of the week came on Friday, when I went to Soros's house in Bedford, New York, north of New York City.

George Soros was eighty-seven years old at the time and exceptionally full of life. I arrived as he was returning from a tennis match. George welcomed me to his home, introduced me to his wife, Tamiko, and begged my forbearance while he took a quick shower before our session. During the short wait, Tamiko quizzed me about the *Washington Monthly* article.

Soros's speech was already great when I first saw it, but he is a perfectionist and thought that it could be much better. We spent more than four hours editing it line by line until George expressed satisfaction with the substance. When we got to the end, I assumed my job was done, but George asked if I could return the following day, which was Saturday. He wanted to review the speech again and then prepare for questions he might get from the press. No one was expecting George Soros to deliver a warning about the dangers of internet platform monopolies, and he wanted to be certain he understood the technical issues described in my essay well enough to answer questions from journalists. On Saturday morning, George, Tamiko, Michael Vachon, and I sat around the dining room table, reviewing every issue from multiple directions until George felt he could respond to whatever the reporters asked. It took more than three hours. The speech is included in the appendices to this book.

Soros's commitment to democracy around the world had particular salience in January 2018, given the rise of Donald Trump and the growing strength of hypernationalists in Europe. After an opening that focused on geopolitics, Soros's speech turned to the threat to democracy from internet monopolies like Google and Facebook. Playing to his strengths, Soros framed the threat in economic terms. He characterized the internet monopolies as extraction businesses in the vein of oil companies, but with a better business model. Network effects allow for ever higher returns as they grow. Growth requires more time and attention from every user every year, gained through surveillance of users and designs that create psychological addiction. Exceptional reach enables monopolies to act as gatekeepers. Media companies must work with them on their terms. Yet the internet monopolies do not acknowledge responsibility for the content on their site, which allows disinformation to flourish. Active users lose the ability to separate fact from fiction, making them vulnerable to manipulation. Soros emphasized the potential threat to democracy of an alliance between authoritarians and internet monopolies. He warned that the monopolies are vulnerable both to China's influence and to competition from Chinese companies playing their version of the same game. Soros concluded the speech with praise for the European Union's approach to protecting users from internet monopolies. George may have started from my *Washington Monthly* essay, but the final speech went much further, tying the threat from internet platforms to geopolitics. By the time I left the Soros home, I had every hope that George's speech would have an impact.

It did. The Soros speech at Davos on January 25 reverberated through the halls of government in both Europe and the United States, reframing the conversation from the relatively narrow confines of the US presidential election to the much broader space of global economics and politics. Policy makers take Soros very seriously, even the ones who disagree with him, and few expected the eight-seven-year old billionaire to speak so thoughtfully and forcefully about technology. For many, it was a wake-up call. In the US, where policy makers' enormous trust in tech platforms had already been tested by the congressional hearings three months earlier, Soros added pressure for a further reassessment. For users, the effect of Soros's speech was more abstract. Most users really like Facebook. They really like Google. There is no other way to explain the huge number of daily users. Few had any awareness of a dark side, that Facebook and Google could be great for them but bad for society. I don't know how many users heard Soros's speech—probably fewer than watched at least some of the congressional hearings back in October and November, which itself was not a big number—but many more saw the headlines. I suspect the details of Soros's argument did not fully register, but coming so soon after the hearings, the headlines had to leave an impression: same companies, new problems. An increasing number of users registered an awareness of controversy surrounding Facebook and Google. More knew that the platforms had issues, even if they did not yet know the particulars or how they might be affected personally.

9

The Pollster

Technology is cool, but you've got to use it
as opposed to letting it use you. —PRINCE

Andrew Bosworth's tweet—"Who the fuck is Roger McNamee?"—was not the kind of engagement I had hoped for from Facebook. It was hard to imagine that Boz viewed me as a threat, so why bother to mention me? As we had learned from Zuck's New Year's resolution, any response at all from Facebook engaged the news media. It amplified our signal, increasing both awareness and skepticism about Facebook's business practices in the context of the 2016 election.

Zuck's New Year's resolution was just the first in a series of public pronouncements. It was followed ten days later by changes to News Feed. Facebook demoted publisher content, while promoting posts from family, friends, and Groups. It positioned the change as an effort to reduce fake news and increase content from the sources users trusted most. Skeptics suggested that Facebook made the change to reduce the risk that regulators would view it as a media company. Less publisher content would also minimize the opportunity for accusations of editorial bias. The problem is that Facebook really is a media company. It exercises editorial judgment in many ways, including through its algorithms. Facebook's position has always been that users choose their friends and which links to view, but in reality, Facebook selects and sequences content for each user's News Feed, an editorial process that had led to criticism in the past, most notably when conservatives accused the company in May 2016 of bias in its Trending Stories feature. At that time, human editors curated Trending Stories. Stories with a conservative slant represented something less than half in the spring of 2016, the result of multiple factors. Facebook may have bowed to the accusation of bias because of cofounder Chris Hughes's role in President Obama's reelection campaign in 2012, when Hughes ran digital operations. Whatever the reason, Facebook's decision in May 2016 to replace human curators with algorithms proved to be a disaster. Far-right voices gamed the algorithm effectively, and disinformation dominated Trending Stories, just in time to amplify the Clinton email server story.

Whether or not the changes to News Feed were another attempt to deny responsibility for third-party content on the platform, they had the effect of promoting the primary elements of filter bubbles—family, friends, and Groups—at the expense of the content most likely to pierce a filter bubble, journalism. The changes seemed like a step backward. Had they been implemented in 2015, they would likely have magnified the effect of the Russian interference.

February 2018 began with two events that were our first efforts to put organizational muscle behind our effort: the launch of the Center for Humane Technology (CHT) and the one-day Truth About Tech conference on kids and social media. CHT is a not-for-profit dedicated to helping consumers deal with the dark side of technology. *The New York Times* wrote a launch story about CHT that included a list of founders and advisors, two of which were Facebook alumni. Both suffered through hostile phone calls from Facebook for associating with us. While the mission of CHT was nonpartisan and user-centric, Zuck and Sheryl did not take kindly to former colleagues lending their names to it. Other than annoying the two Facebook alumni, Facebook's reaction had no impact on CHT. This was confirmed the following day, when we convened in Washington for the Truth About Tech conference. Explicitly positioned as a joint project of Common Sense Media and the Center for Humane Technology, Truth About Tech had a speaker list that included Senators Mark Warner and Edward Markey, as well as Representative John Delaney; Dr. Robert Lustig, the pediatric endocrinologist who had exposed the addictive properties of sugar; author Franklin Foer; Chelsea Clinton; and Tristan, Randima Fernando, and me from the CHT team. Randima had recently joined CHT, after working with Tristan on Time Well Spent. Common Sense Media had invited people from Facebook and Google to attend the conference, hoping for constructive dialogue about protecting children from the harmful aspects of screens and online content, but they were not willing to engage in the conversation.

At the kickoff event for the conference, House Minority Leader Nancy Pelosi sought me out. We have similar taste in music and had shaken hands a few times backstage at Grateful Dead and U2 concerts. She took me aside, thanked me for the work our team had been doing with House Intelligence, and asked if we needed help reaching other members of the Democratic House caucus. The shortest unit of time in the universe is how long it took me to say yes. She recommended that I brief her entire staff as a first step, which I did a few weeks later. That momentary encounter would soon pay larger dividends.

That same day, *The Verge* published a story by Casey Newton about Tavis McGinn, who had recently left Facebook after a six-month stint as the personal pollster for Zuck and Sheryl. The story shocked us. Why would Facebook—which employs a small army to survey users on every issue imaginable—need to hire a new person for the sole purpose of polling the popularity of its two top executives? More remarkable was the timing: Tavis had been at Facebook from April through September 2017. They had hired the pollster while they were still denying any involvement in the Russian interference.

In the article, Tavis explained that his experience at Facebook did not work out the way he had hoped.

"I joined Facebook hoping to have an impact from the inside," he says. "I thought, here's this huge machine that has a tremendous influence on society, and there's nothing I can do as an outsider. But if I join the company, and I'm regularly taking the pulse of Americans to Mark, maybe, just maybe that could change the way the company does business. I worked there for six months and I realized that even on the inside, I was not going to be able to change the way that the company does business. I couldn't change the values. I couldn't change the culture. I was probably far too optimistic.

"Facebook is Mark, and Mark is Facebook," McGinn says. "Mark has 60 percent voting rights for Facebook. So you have one individual, 33 years old, who has basically full control of the experience of 2 billion people around the world. That's unprecedented. Even the president of the United States has checks and balances. At Facebook, it's really this one person."

No one on our team had interacted with Zuck and Sheryl since the 2016 election. We could only guess what they were thinking. Here was someone who knew what they were thinking just a few months earlier. We were keen to connect with him. A reporter at *The Washington Post*, Elizabeth Dwoskin, knew how to reach Tavis and volunteered to introduce us. It took a couple of days, but Tavis called my cell phone while I was riding the subway in New York. Fortunately, the call came through while my subway train was stopped at the Twenty-eighth Street station on the 1 line. I hopped off the train and spoke to Tavis for nearly half an hour on the subway platform, our conversation punctuated by the noise of passing trains.

Tavis reframed my understanding of the psychology of Zuck and Sheryl, as well as the way that Facebook's culture had evolved since the IPO. He emphasized that spectacular success had cemented Zuck's position as the undisputed leader of the world's largest network of people. As Zuck's partner, Sheryl enjoyed comparable regard from employees. Internally, everyone defers to the Big Two. Tavis's hiring reflected Zuck's and Sheryl's concern for their personal brands, which they feared would be tarnished by any negative feedback that came Facebook's way. Tavis was convinced that both Zuck and Sheryl had bigger plans after Facebook, and those plans were threatened by the rising tide of criticism. Soon after joining the company, Tavis learned that neither Zuck nor Sheryl wanted to hear bad news. As much as I would have liked to dig deeper, I respected Tavis for honoring Facebook's nondisclosure agreement (NDA).

On February 16, Special Counsel Robert Mueller issued a thirty-seven-page indictment of thirteen Russian nationals and three organizations for interference in the 2016 US election, wire fraud, and bank fraud. The indictment referred to Facebook, Instagram, and Twitter by name and underscored the ease with which the Russians had exploited the architecture and algorithms of social platforms to spread disinformation and suppress votes. The story landed like a bombshell. As a bonus, the world got a peek into Facebook's culture, thanks to a tweet storm from Rob Goldman, Facebook's vice president of advertising, in response to the indictments.

The president of the United States retweeted Goldman's tweet, making it a global news story with consequences for which Facebook had not prepared.

Rob Goldman ✔
@robjective

(Follow) ∨

Very excited to see the Mueller
indictment today. We shared Russian
ads with Congress, Mueller and the
American people to help the public
understand how the Russians abused our
system. Still, there are keys facts about
the Russian actions that are still not well
understood.

5:57 PM - 16 Feb 2018

Rob Goldman ✔
@robjective

(Follow) ∨

Most of the coverage of Russian
meddling involves their attempt to effect
the outcome of the 2016 US election. I
have seen all of the Russian ads and I
can say very definitively that swaying the
election was *NOT* the main goal.

5:57 PM - 16 Feb 2018

Rob Goldman ✔
@robjective

(Follow) ∨

The majority of the Russian ad spend
happened AFTER the election. We
shared that fact, but very few outlets
have covered it because it doesn't align
with the main media narrative of Tump
and the election.

Hard Questions: Russian Ads Delivered to Congre...
What was in the ads you shared with Congress? How
many people saw them?
newsroom.fb.com

5:57 PM - 16 Feb 2018

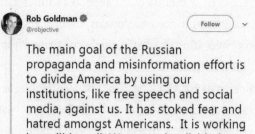

Rob Goldman ✅
@robjective

Follow ⌄

The main goal of the Russian propaganda and misinformation effort is to divide America by using our institutions, like free speech and social media, against us. It has stoked fear and hatred amongst Americans. It is working incredibly well. We are quite divided as a nation.

5:57 PM - 16 Feb 2018

Rob Goldman ✅
@robjective

Follow ⌄

The single best demonstration of Russia's true motives is the Houston anti-islamic protest. Americans were literally puppeted into the streets by trolls who organized both the sides of protest.

Russian Trolls Organized Both Sides of an Islam Protest in Texas
Update: Nov. 3 — Houston counter-protesters are alleging that their protest was not connected to the Russian-led group. The story now reflects those...
sacurrent.com

5:57 PM - 16 Feb 2018

Goldman's tweets provided a glimpse into the state of mind of Facebook executives. Apparently, they viewed the Mueller indictment as exoneration. They thought it was okay that the Russians had exploited Facebook and Instagram to "divide America," so long as they did not intend to sway the election. It demonstrated a stunning lack of situational awareness by a member of Facebook's inner circle. Journalists and the blogosphere mocked Goldman and Facebook for being clueless. Goldman's tweet storm had effectively confirmed Tavis's view of the mind-set inside Facebook.

A few days later, I met Tavis for the first time, in a coffeehouse on Market Street in San Francisco. He told me that he hailed from North Carolina and went to college at the University of North Carolina at Chapel Hill, where he discovered a talent for entrepreneurship, launching a business that rented refrigerators to students. After college, Tavis began his career in market research. He became an expert at creating and executing surveys, which eventually landed him at the insurance company GEICO, then at Google. Facebook pursued him several times at Google, but Tavis was not tempted until early 2017, when the opportunity arose to be the personal pollster for Zuck and Sheryl.

With the Mueller indictment and the subsequent Goldman tweet storm as context, Tavis and I shared insights for ninety minutes. After leaving Facebook, Tavis had launched a new market research business, Honest Data, but hoped to join our team on a part-time basis, with the goal of communicating his concerns about Facebook to regulators. He had done research on the NDA that every Facebook employee must sign and concluded that it could be pierced in the event of a legal case, such as an investigation by a state attorney general. While we were not actively working with a state attorney general, we had built a relationship with the attorney general of New York and members of his team. I offered to make an introduction. It made sense to initiate relationships with other attorneys general.

Two days later, on Wednesday, Tristan and I went to Seattle to meet the chief of staff to the attorney general of Washington, the adolescent mental health team at the Gates Foundation, and the CEO of Microsoft. The three had very different interests and no prior exposure to us, so the best we could hope for would be to plant seeds. We did that in the first two meetings, before going to Microsoft headquarters for a meeting with the CEO, Satya Nadella, and the head of business development, Peggy Johnson. Nadella's book, *Hit Refresh*, had expressed a philosophy consistent with the values Tristan had set for the Center for Humane Technology, and we hoped to secure his support for our effort. Companies like Microsoft don't commit in the first meeting, but Satya and Peggy engaged thoughtfully with us. Satya noted that while two Microsoft products, Xbox and LinkedIn, used techniques from the Fogg playbook, most of its products did not. There would be significant benefits to adopting Tristan's notions of humane design in the core Windows line. He asked that Tristan schedule a return visit to brief Microsoft's engineering leadership. On the way out of Satya's office, we ran into Microsoft founder Bill Gates, on his way to visit Nadella. In an Axios interview a week earlier, Bill had criticized the tech giants of Silicon Valley for acting cavalierly, suggesting that they risked the kind of government regulation that had plagued Microsoft. "The companies need to be careful that they're not . . . advocating things that would prevent government from being able to, under appropriate review, perform the type of functions that we've come to count on," he warned.

Even apart from the Mueller indictment and Goldman's tweets, February delivered a series of public relations failures to Facebook. Two huge consumer packaged goods vendors, Procter & Gamble and Unilever, criticized Facebook (and Google) and threatened to stop advertising. P&G expressed displeasure with the platforms' lack of transparency and accountability and argued that advertisers should not tolerate the biggest internet platforms' failure to conform to the disclosure standards of the advertising industry. Current practices caused P&G to ask if it was getting what it paid for. Unilever objected to fake news, extremist content, and the role that the platforms had played in sowing discord.

Days later, journalists disclosed that Facebook had sent millions of marketing messages to phone numbers that users had provided as part of a security feature called two-factor authentication, something it had promised not to do. A storm of criticism followed, eliciting a couple of tweets from Facebook's vice president of security, Alex Stamos, that reprised Facebook's tin ear to legitimate criticism. Once again, a press revelation confirmed Tavis's hypothesis about the company's culture. Next, a court in Belgium ruled that Facebook broke privacy laws and ordered the company to stop collecting user data in that country. Even when news reports dealt with another tech platform—as was the case when journalists revealed that Russian bots promoted disinformation on Twitter in the wake of the school shooting in Parkland, Florida—Facebook's reputation took a hit. Finally, a virtual-reality shooting game that Facebook sponsored at the Conservative Political Action Conference (CPAC) drew massive criticism, also coming on the heels of the Parkland shooting.

It seemed as though a negative story hit Facebook almost every day. In a news environment dominated by the unprecedented behavior of the Trump administration, the Facebook story kept breaking through. Most of the world was barely paying attention to the details, but the story had legs.

Facebook continued to defend its business model. In the past, apologies had always been enough to neutralize criticism. Not this time. Ignoring criticism for more than a year had not worked. Zuck's 2018 New Year's resolution and the tweets from inner-circle executives had backfired, but with those responses, Facebook acknowledged the criticism for the first time. Public pressure made a difference.

MY ORIGINAL FEAR—that Facebook's problems were systemic—had been validated repeatedly by journalists, policy makers, and the

Mueller investigation. Facebook executives had begun to engage, but it seemed that the rest of the company continued to operate as it always had. We heard rumors of internal discontent, but that may have been wishful thinking. No new whistle-blowers had emerged, and no one inside the company leaked any data to support the investigations. But the pressure on Facebook was about to intensify.

10

Cambridge Analytica Changes Everything

Once a new technology rolls over you, if you're not part of
the steamroller, you're part of the road. —STEWART BRAND

March 2018 brought almost daily revelations about unintended damage from social media. *Science* magazine published a study conducted by professors at MIT of every controversial story in English on Twitter. It revealed that disinformation and fake news are shared 70 percent more often than factual stories and spread roughly six times faster. The study noted that bots share facts and disinformation roughly equally, suggesting that it is humans who prefer to share falsehoods. No one claimed that the problem might be confined to Twitter. The study provided further evidence that the dark side of social networks may be systemic, driven by design choices that favor some of the worst aspects of human behavior.

As if on cue, disinformation spread by Infowars about bombing suspects in Austin, Texas, reached the top of the charts on YouTube. YouTube responded to this failure in moderation by trying to pass the buck to Wikipedia, which it claimed would debunk disinformation. When I checked with Katherine Maher, the executive director of Wikipedia, that day, I learned that YouTube had made the announcement without first speaking to or offering to reimburse Wikipedia, a nonprofit with a small professional staff. When Wikipedia pushed back, the folks from YouTube seemed not to understand why Wikipedia would not want to spend its limited budget being YouTube's fact-checker.

For Facebook, a tide of bad news washed in from outside the United States. Early in March, the Sri Lankan government ordered internet service providers to block Facebook, Instagram, and WhatsApp temporarily, due to an explosion of real-life violence against that country's Muslim minority, triggered by online hate speech. The government criticized Facebook and its subsidiaries for not taking action to limit hate speech on their platforms, a criticism that would increasingly be echoed in jurisdictions around the world. Facebook's response—that its Community Standards prohibited the inciting of violence—was classic. In Sri Lanka, the Community Standards were in English, spoken by 24 percent of the population, rather than Sinhala, which is spoken by 87 percent. As in other countries, Facebook did not prioritize the enforcement of its Community Standards.

Days later, a United Nations report accused Facebook of enabling religious persecution and ethnic cleansing of the Rohingya minority in Myanmar. As in Sri Lanka, hate speech on Facebook triggered physical violence against innocent victims. According to Médecins Sans Frontières, the death toll from August through December 2017 was at least nine thousand. In a country where Facebook completely dominates social media, the platform plays a central role in communications. *The Guardian* reported:

> The UN Myanmar investigator Yanghee Lee said Facebook was
> a huge part of public, civil and private life, and the government
> used it to disseminate information to the public.

"Everything is done through Facebook in Myanmar," she told reporters, adding that Facebook had helped the impoverished country but had also been used to spread hate speech.

"It was used to convey public messages but we know that the ultra-nationalist Buddhists have their own Facebooks and are really inciting a lot of violence and a lot of hatred against the Rohingya or other ethnic minorities," she said.

"I'm afraid that Facebook has now turned into a beast, and not what it originally intended."

Slate quoted Facebook executive Adam Mosseri as saying the situation in Myanmar was "deeply concerning" and "challenging for us for a number of reasons." He argued that in Myanmar, Facebook had not been able to employ its standard practice of working with third-party fact-checkers. Instead, it was attempting to regulate hate speech through its terms of service and community standards, which presumably are at least as opaque to users in countries like Myanmar as they are in the United States. And the situation is actually worse than that. Facebook has a program called Free Basics, which is designed for countries where mobile communications is available but too expensive to permit widespread use of internet services. The functionality of Free Basics is limited, designed to plant seeds in developing countries, while gaining positive press for doing so. Sixty emerging countries around the world have embraced Free Basics, many with minimal experience with telecom and media, and some have been disrupted by it. In Myanmar, Free Basics transformed internet access by making it available to the masses. That is also the case in most of the other countries that have adopted the service. The citizens of these countries are not used to getting information from media. Prior to Free Basics, they had little, if any, exposure to journalism and no preparation for social media. Their citizens did not have filters for the kind of disinformation shared on internet platforms. An idea that sounded worthy to people in the US, Free

Basics has been more dangerous than I suspect its creators would have imagined.

In Myanmar, a change in government policy caused an explosion in wireless usage, making Facebook the most important communications platform in the country. When allies of the ruling party used Facebook to promote violence against the Rohingya minority, the company fell back on its usual strategy of an apology and a promise to do better. The truth is that countries like Myanmar are strategically important to Facebook, but only as long as the cost of doing business there remains low. Facebook had not hired enough employees with the language skills and cultural sensitivity to avoid disruption in countries like Myanmar or Sri Lanka. The company demonstrated little urgency to address this issue, framing it as a process problem rather than a humanitarian crisis.

On March 16, all hell broke loose.

It began when Facebook announced the suspension of a political consulting firm, Cambridge Analytica, and its parent, SCL Group, from the platform. This turned out to be an attempt to preempt a huge story that broke the following day in two newspapers in the United Kingdom, *The Observer* and *The Guardian*, as well as in *The New York Times*. The *Guardian* story opened with a bang:

> The data analytics firm that worked with Donald Trump's election team and the winning Brexit campaign harvested millions of Facebook profiles of US voters, in one of the tech giant's biggest ever data breaches, and used them to build a powerful software program to predict and influence choices at the ballot box.
>
> A whistleblower has revealed to the *Observer* how Cambridge Analytica—a company owned by the hedge fund billionaire Robert Mercer, and headed at the time by Trump's key adviser Steve Bannon—used personal information taken without authorisation in early 2014 to build a system that could profile

individual US voters, in order to target them with personalised political advertisements.

Christopher Wylie, who worked with a Cambridge University academic to obtain the data, told the *Observer*: "We exploited Facebook to harvest millions of people's profiles. And built models to exploit what we knew about them and target their inner demons. That was the basis the entire company was built on."

The story suggested that Cambridge Analytica had exploited a researcher at Cambridge University, Aleksandr Kogan, to harvest and misappropriate fifty million user profiles from Facebook. Kogan, a researcher who was also affiliated with a university in St. Petersburg, Russia, had previously worked on research projects with Facebook. Cambridge University had originally rejected Kogan's request to access its data, leading Kogan and his partner, Joseph Chancellor, to start a company, funded by Cambridge Analytica, that would create a new data set of American voters. They created a personality test that would target Facebook users and recruited test takers from an Amazon service that provided low-cost labor for repetitive information technology projects. Two hundred seventy thousand people were paid one to two dollars each to take the test, which was designed to collect the personality traits of the test taker, as well as data about friends and their Facebook activities. The people who took the test had to be American, and it turned out that they had a lot of friends, more than forty-nine million of them.

Cambridge Analytica was created in 2014 as an affiliate to SCL Group, a British firm that specialized in market research based on psychographics, a technique designed to categorize consumers according to personality types that might have predictive value in the context of elections. In the world of market research, there is considerable doubt about how well psychographics work in their current form, but that issue did not prevent Cambridge Analytica from finding clients, mostly on the far right. To serve the US market, SCL needed to obey federal election laws. It created a US affiliate staffed by US citizens and legal residents. Reports indicated that Cambridge Analytica took a casual approach to regulations. The team of Robert Mercer and Steve Bannon financed and organized Cambridge Analytica, with Alexander Nix as CEO. The plan was to get into the market within a few months, test capabilities during the 2014 US midterm elections, and, if successful, transform American politics in 2016. To be confident that their models would work, Nix and his team needed a ton of data. They needed to create a giant data set of US voters in a matter of months and turned to Kogan to get one. According to Wylie, the Kogan data set formed the foundation of Cambridge Analytica's business. Cambridge Analytica's election-centric focus clearly violated Facebook's terms of service, which did not permit commercial uses of Kogan's data set, but Wylie reported that Facebook made no attempt to verify that Kogan had complied.

At the time that Kogan and Cambridge Analytica misappropriated fifty million user profiles, Facebook was operating under a 2011 consent decree with the FTC that barred Facebook from deceptive practices with respect to user privacy. The decree required explicit, informed consent from users before Facebook could share their data. Apparently, Facebook had taken no steps to secure consent from the friends of the 270,000 test takers, which is to say, something like 49.7 million Facebook users. While the enforcement language of the consent decree is ambiguous, its intent is clear: Facebook had an obligation to safeguard the privacy of its users and their data. You could almost sense a shock wave as people understood the ease with which Kogan had harvested fifty million profiles. Facebook made it easy.

Speculation by journalists and pundits about legal issues that might arise from the Cambridge Analytica story lit up Twitter for hours.

Legal analysts focused on the possibility of a data breach that might have placed Facebook in violation of state laws and FTC regulations. Failure to comply with the FTC consent decree carried a penalty as high as forty thousand dollars per offense. In the case of Cambridge Analytica, the penalty could potentially be measured in the trillions of dollars, far more than the value of Facebook. Cambridge Analytica might be vulnerable to prosecution for fraud and campaign finance violations.

The Cambridge Analytica story transformed the conversation about Facebook, providing something to worry about for just about everyone. Those troubled by the role of Facebook in the 2016 US presidential election could obsess over Facebook enabling inappropriate access to user profiles and how that might have affected the outcome. Those worried about privacy on Facebook saw their worst fears validated. Kogan harvested fifty million user profiles under a Facebook program designed to let third-party app vendors gain access to friends lists. How many third-party app vendors had taken advantage of the program? Only a tiny fraction of affected users knew their profile had been harvested. The data was still out there—quite likely still available for use—with no way to get it back.

Facebook tried and failed to minimize fallout from the story. Having failed to preempt it on day one, Facebook tried to reframe the story in a way that shifted all the blame to Cambridge Analytica. Facebook's initial response, a series of tweets from vice president of security Alex Stamos, denied *The Guardian*'s characterization that Kogan and Cambridge Analytica had committed a data breach. Stamos emphasized that Kogan had authorization to harvest friends lists for research purposes and that the guilty party was Cambridge Analytica, which had misappropriated the user profiles. According to Stamos's tweets, Facebook was a victim.

In disputing the initial characterization of a "data breach," Facebook inadvertently made its public relations problem worse, so much so that Stamos deleted the tweets. When it described Kogan as a legitimate researcher, Facebook effectively acknowledged that the harvesting of user profiles by third parties was routine. Our team hypothesized that every Facebook user profile in that era was harvested at least once. Facebook's acknowledgment came as a shock. It should not have been. We soon learned that sharing private user data with third parties was one of the core tactics that contributed to Facebook's success.

In its early years, Facebook had been far more successful in growing its user base than in growing the amount of time each user spent on the site. The introduction of third-party games, particularly Zynga's Farm-Ville in 2009, changed that by rewarding social interaction and leveraging friends lists to boost the population of players. By March 2010, FarmVille had more than 83 million monthly users and 34.5 million daily users, and it changed the economics of Facebook. Zynga leveraged its user growth with in-game advertising and purchases, which translated rapidly into revenue of hundreds of millions of dollars. On Facebook, 30 percent of Zynga's in-game advertising and purchase revenues went to Facebook, making Zynga a key partner in the era before Facebook had a scalable advertising business model. In the year prior to the IPO, Zynga alone accounted for twelve percent of Facebook's revenue. Zynga's ability to leverage friends lists contributed to an insight: giving third-party developers access to friends lists would be a huge positive for Facebook's business. Social games like FarmVille cause people to spend much more time on Facebook. Users see a lot of ads. Zynga had a brilliant insight: adding a social component to its games would leverage Facebook's architecture and generate far more revenue, creating an irresistible incentive for Facebook to cooperate. In 2010, Facebook introduced a tool that enabled third-party developers to harvest friends lists and data from users. They saw the upside of sharing friends lists. If they recognized the potential for harm, they did not act on it. Despite the 2011 consent decree with the FTC, the tool remained available for several more years.

Kogan's data set included not only Facebook user IDs but also a range of other data, including activity on the site. Such a list had great value if used inside Facebook, but Cambridge Analytica had bigger plans. They married the data set to US voter files, which include both demographic information and voting history. According to Wylie, Cambridge Analytica was able to match at least thirty million Facebook profiles to voter files, equivalent to 13 percent of all eligible voters in the country. At that scale, the data set had tremendous value to any campaign. Facebook's advertising tools allow targeting by demographics and interests but are otherwise anonymous. Tying the voter files to the user profiles would have enabled Cambridge Analytica to target advertising inside Facebook with exceptional precision, particularly if one of the goals was voter suppression. In 2016, the winner in the electoral college lost the popular tally by nearly three million votes. Three states, which Trump won by a total of 77,744 votes, provided more than the margin of victory in the electoral college. Is it possible that the Cambridge Analytica data set might have influenced the outcome? Yes. It's virtually impossible that it didn't.

Targeting inside Facebook mattered because it works. Cambridge Analytica's original client, the presidential campaign of Senator Ted Cruz, complained that the psychographic models sold by Cambridge Analytica did not work for them. In the end, psychographics probably didn't matter to the Trump campaign. They had more powerful weapons available to them, in the form of Cambridge Analytica's data set of thirty million enhanced voter files and Facebook's targeting tools and employees.

After the initial bombshell story, *The Guardian* published a video interview with the whistle-blower, Christopher Wylie, whose pink hair made him instantly recognizable and ubiquitous online and in newspapers. The UK's ITN Channel 4 added to the story with a series of undercover exposés about Cambridge Analytica that reflected very badly on that company and, by extension, on Facebook. In one of the exposés, senior executives of Cambridge Analytica are captured on film bragging about their ability to use prostitutes to entrap politicians.

The Guardian also reminded readers that it had previously revealed the connection between Kogan and Cambridge Analytica in a story in December 2015. Facebook claimed at the time not to have known that Cambridge Analytica had gained possession of Kogan's data set. Citing a violation of its terms of service, Facebook sent letters to Cambridge Analytica and Kogan, insisting that they destroy all copies of the data set and certify that they had done so by checking off a box in a form. Facebook never audited either Cambridge Analytica or Kogan and did not dispatch inspectors to confirm destruction of the data set. Once again, the focus at Facebook was on protecting against legal liability, not on protecting users.

Facebook's argument that it had been a victim of Cambridge Analytica fell apart when *Slate*'s April Glaser reminded her readers that the company had hired and continued to employ Joseph Chancellor, who had been Aleksandr Kogan's partner in the startup that harvested Facebook user profiles on behalf of Cambridge Analytica. Facebook had known about the connection between Cambridge Analytica and Kogan/Chancellor since at least December 2015. They should have been really angry at Kogan and Chancellor for misappropriating the data set. Why would they hire someone who had misappropriated private user data? And yet Chancellor was now a Facebook employee. The Glaser story was actually old news that acquired new salience in the context of recent revelations. The Facebook/Kogan/Chancellor link had originally been reported by *The Intercept* in March 2017, and it had connected the dots from Cambridge Analytica to Kogan to Chancellor to Facebook in a way that did not make anyone look good. Chancellor no longer works at Facebook.

If the relationship between Facebook, Kogan, and Cambridge Analytica had been known since late 2015, why was this story a much bigger deal the second time around? The short answer is that context had been missing the first time the story made news. Unlike in December 2015, we now knew that the Russians had exploited Facebook to sow discord among Americans and then support Donald Trump's presidential candidacy. We also knew that Cambridge Analytica had been the Trump campaign's primary advisor for digital operations and that Facebook had embedded three employees in the Trump campaign to support that effort. The presidential election had been especially close, and it seemed likely that Trump's targeting of voters in key states late in the campaign had been decisive. There were half a dozen things that had to happen on election day in order for Trump to win, and successful Facebook advertising in key states was one of them. The new context made it hard to escape the conclusion that Cambridge Analytica and the Trump campaign had exploited Facebook, just as the Russians had. Little had been learned about Facebook's engagement with Russian agents, but there could be little doubt that Facebook had willingly engaged with Kogan, Cambridge Analytica, and the Trump campaign. It was entirely possible that Facebook employees had played a direct role in the success of Trump's digital strategy on Facebook.

It is hard to overstate the impact of the Cambridge Analytica story. Coming on the heels of so many other bad stories, it confirmed many people's worst fears about Facebook. The relentless pursuit of growth had led Facebook to disregard moral obligations to users, with potentially decisive consequences in a presidential election and as yet unknown other consequences to the millions of users whose data had been shared without prior consent. The national conversation Tristan and I had hoped to start eleven months earlier had reached a new level. The next few days were going to be a real test for Facebook. There was no good way to spin the Cambridge Analytica story. How would they handle the reckoning?

The first wave of stories about Cambridge Analytica had an especially profound effect on one member of our team, Sandy Parakilas. Since joining us in November 2017, Sandy had dedicated himself to our cause, using his experience from Facebook to shed light on the dark side of social media in op-eds and interviews. He had also played a role in the rollout of the Center for Humane Technology. Overnight, Sandy's role changed dramatically. He went from activist to whistle-blower. From 2011 to 2012, Sandy had been an operations manager for Facebook Platform, home to all third-party applications on the site. Sandy had a uniquely valuable perspective on Facebook's policies and actions relative to user data privacy and security.

The child of academics in Maine, Sandy grew up wanting to be a jazz drummer. He gave that career a try, then opted for business school in pursuit of a more stable path. Sandy won a national business plan competition to recast the business model of a not-for-profit called One Laptop per Child. That success helped Sandy secure an offer from Facebook, which evolved into a newly created position focused on user privacy. In classic Facebook style, the company installed an inexperienced and untested recent graduate in a position of great responsibility, a position to which other companies would have traditionally assigned someone with meaningful relevant experience. The job did not contribute to growth, which meant it would not be a high priority inside Facebook. In fact, the task of protecting user privacy would internally be viewed as a form of friction—which put Sandy in a very difficult position. To have any chance of success, he had to push back against a core element of Facebook's culture and make the company embrace at least one form of friction—privacy protection—that almost certainly would have at least a small negative impact on growth. It did not take long for Sandy to understand that the odds were against him.

In November 2011, early in Sandy's tenure, Facebook entered into its consent decree with the FTC to settle an eight-count complaint about material misrepresentations to users with respect to privacy. The complaint included a long list of Facebook misrepresentations, beginning in 2009. In some cases, Facebook promised one thing and did the exact opposite. In others, Facebook made commitments to regulators that it did not keep. According to the FTC's press release:

> The proposed settlement bars Facebook from making any further deceptive privacy claims, requires that the company get consumers' approval before it changes the way it shares their data, and requires that it obtain periodic assessments of its privacy practices by independent, third-party auditors for the next 20 years.

The spirit of the settlement is obvious, and the original deal with Aleksandr Kogan was one of many that appeared to violate it.

With respect to the sharing of user data, the consent decree seemed to offer Facebook two options: it could eliminate the tool for harvesting friends lists and/or it could create a team to enforce the consent decree by monitoring and auditing third-party developers. According to Sandy, at the time Facebook did neither. Sandy's requests for engineering resources to enforce the decree were denied with an exhortation to "figure it out." In the end, the consent decree's enforcement provision gave Facebook a "get out of jail free" card. The FTC allowed Facebook to both pick and pay the third-party auditor whose certification of compliance with the consent decree would be required. Facebook did not have to worry about compliance. It received passing grades every time, even as it failed to comply with the spirit of the decree.

At Facebook, figuring it out is a way of life. The company got its start with a bunch of Harvard undergraduates who knew how to code but had almost no experience with anything else. They figured it out. Each new wave of employees followed the same path. Some took too long and were pushed out. The rest got comfortable with the notion that experience was not helpful. At Facebook, the winners were people who could solve any problem they encountered. The downside of this model is that it encourages employees to circumvent anything inconvenient or hard to fix.

In the run-up to Facebook's May 2012 IPO, the company scrutinized every aspect of its business. A series of privacy issues emerged that related to Facebook Platform, specifically to the tool that enabled third-party apps to harvest data from users' friends. As Sandy describes it, Facebook's lack of commitment to user data privacy created issues of disclosure and legal liability that could and should have been addressed before the initial public offering. That did not happen. Recognizing that Facebook did not intend to enforce the spirit of the FTC consent decree—and would blame him if ever there was bad press about it—Sandy quit his job.

Two years later, the person who occupied Sandy's position as operations manager for Facebook Platform approved Aleksandr Kogan's application to use a personality test to harvest friends data for academic research. Eighteen months after that, the person in that position sent a letter to Kogan and Cambridge Analytica in response to the article in *The Guardian*, ordering them to destroy the data set and certify that they had done so. While I have no proof one way or the other, I hypothesize that Facebook's commitment to enforcing the spirit of the FTC consent decree was no greater in 2014 than it was during Sandy's time at the company. If it had been greater, they almost certainly would have exercised their right to audit and inspect Kogan and Cambridge Analytica to ensure compliance.

The Cambridge Analytica story caused our team to rethink everything we knew about two things: the number of people who might have been affected by Facebook's casual approach to privacy and the company's role in the 2016 presidential election. On privacy, we had one terrifying data point: Facebook disclosed that there were nine million applications on Facebook Platform at the time of the IPO in 2012. In theory, all of them might have attempted to use the tool for friend harvesting. We knew that wasn't true because many of the "applications" were single pages created by third parties, designed to communicate a message rather than gather data. If only 1 percent of the apps on Platform harvested friends data, however, that would still have been ninety thousand applications. Sandy confirmed that during his tenure, the number of apps that harvested data was in the tens of thousands. The program continued for two years after Sandy left, so the number may have grown.

And as bad as the Cambridge Analytica harvest was, it was nowhere near the largest. The problem, as Sandy described it, was that some of the applications that harvested data were huge. Games like CityVille and Candy Crush had peak user bases that approached one hundred million players worldwide, including perhaps a quarter of all users in the United States. If games like CityVille and Candy Crush harvested friends data, they would have captured nearly every user in the US many times over. Lots of applications had one million users, and every one would have had access to the friends lists of four times as many Facebook users as Cambridge Analytica. The odds that any Facebook user in the 2010–14 period escaped data harvesting are vanishingly small.

Other than a handful of tweets from executives like Alex Stamos, Facebook kept quiet for five days after the Cambridge Analytica story broke. The only news from Facebook also related to Stamos, who announced that he planned to leave the company in five months. Journalists and industry people speculated about whether Stamos was being fired or was leaving voluntarily, but he departed Facebook on August 17 to join the faculty at Stanford. There were several data points to suggest that Stamos had pushed for greater transparency with respect to Facebook's role in the 2016 election but had been overruled by Sheryl Sandberg and Elliot Schrage. A few journalists noted that Stamos's own history made him less sympathetic than he might have been. For example, he led Yahoo's security team when that company built a tool to scan all incoming email on behalf of US intelligence agencies. No internet platform had ever acceded to such a broad request before, and Yahoo was heavily criticized.

When he finally broke his silence after five days, Zuck apologized for "a breach of trust between Facebook and the people who share their data with us and expect us to protect it." As reported in *The Guardian*, Zuck sounded remorseful.

"We have a responsibility to protect your data, and if we can't then we don't deserve to serve you," Zuckerberg wrote. He noted that the company has already changed some of the rules that enabled the breach, but added: "We also made mistakes, there's more to do, and we need to step up and do it."

Zuck promised to change Facebook's rules for sharing data with third-party applications, a promise that rang hollow because the company had eliminated the tool for harvesting friends data in 2014 and could not recapture harvested data. Once the profiles left Facebook, they could have gone anywhere. They could have been copied over and over. And no one knew where the data sets and copies had gone. Could a copy of the Cambridge Analytica data set have found its way to Russian groups like the Internet Research Agency?

In the context of Sandy's experience, one conclusion was inescapable: Facebook had not protected user data privacy because sharing data broadly was much better for its business. Third-party applications increased usage of Facebook—time on site—a key driver of revenue and profits. The more time a user spends on Facebook, the more ads he or she will see and the more valuable that user will be. From Facebook's perspective, anything that increases usage is good. They seemingly never considered the possibility that what they were doing was wrong.

Journalists and policy makers expressed outrage at Facebook's failure to respond formally to the story for five days. Members of Congress and the UK Parliament called for Zuck to testify under oath. The analyst in me could not help but notice that the story of Facebook's role in the 2016 election had unfolded with one consistent pattern: Facebook would first deny, then delay, then deflect, then dissemble. Only when the truth was unavoidable did Facebook admit to its role and apologize. Suddenly, a lot of people understood that apologies had been a standard part of Facebook's public relations tool kit from Zuck's days at Harvard. Zeynep Tufekci, a brilliant scholar from the University of North Carolina, framed Facebook's history as a "fourteen-year apology tour." I reflected that it might be time to tweak Facebook's corporate motto:

Move fast, break things, apologize, repeat.

Zuck embarked on a charm offensive, beginning with interviews with *The New York Times*, CNN, and the tech blog *Recode*. As I watched and read the interviews, I could see why Zuck had waited five days before speaking. He wanted to be prepared. He must have concluded that the damage from bad answers would be greater than that from five days of silence. There is no way to rerun the experiment, but at the time, it seemed that waiting five days increased the brand damage to Facebook, Zuck, and Sheryl; and Zuck's performance in the interviews did not reduce the pressure on Facebook.

The following day, Sheryl began her own apology tour. As the former chief of staff to the secretary of the Treasury, Sheryl has years of experience at the highest levels of politics. As the chief operating officer of Facebook and a board member of the Walt Disney Company, she has been exposed to the widest possible range of business executives and situations. As a bestselling author, she has built a brand with consumers. In all these contexts, she has demonstrated exceptional communications skills. Back in the fall of 2017, in the first interview she did about Russia's use of Facebook to interfere in the 2016 election, Sheryl gave a master class in crisis management communications. She looked and sounded sincere, she was convincing, but when I reviewed the transcript, it was obvious she had not admitted anything substantive or committed Facebook to any material change. This made her failure in the first post–Cambridge Analytica interviews shocking. As Zuck had done, Sheryl seemed to choose interviewers who might not probe deeply. It didn't help. She left a bad impression. Several people who knew Sheryl only by her excellent reputation expressed surprise that she came across so poorly, so lacking in sincerity. Industry professionals responded in one of two ways. Most seemed to want the controversy to go away so they could go back to making money. The relatively small number who were troubled by Facebook's behavior expressed shock that Sheryl could be so unconvincing.

Sheryl Sandberg knows she is capable of achieving any goal to which she commits herself. People who know Sheryl well cannot help but agree. Her talents are exceptional. Ever since her career began, Sheryl has carefully cultivated her public image, controlling every aspect of it, moving from government to business to philanthropy to family. Sheryl's brilliant career did not come without collateral damage to some unfortunate people in her orbit, but in the culture of Silicon Valley, such damage comes with the territory. The trick is to appear virtuous when you take advantage of others. Did Sheryl have plans after Facebook that might suffer from the escalating PR crisis? Where another executive might have leaned into the problem, taking a short-term hit in the hope of long-term benefit, Sheryl had chosen to lean out. She seemed to be avoiding the spotlight when it mattered most. A few industry people and journalists I encountered appeared to be enjoying a moment of schadenfreude.

Even after the first salvo of press appearances by Zuck and Sheryl, the pressure on Facebook continued to grow. On March 21, a Facebook user filed a proposed class action lawsuit in San Jose, California. That same day, this showed up on Twitter:

God   ∨
@TheTweetOfGod

Mark Zuckerberg is one of the last people you should trust, and I mean that both literally and alphabetically.

1:42 PM · Mar 21, 2018

12,905 Retweets   **50,321** Likes

On March 22, a game designer by the name of Ian Bogost published a piece in *The Atlantic* titled, "My Cow Game Extracted Your Facebook Data."

For a spell during 2010 and 2011, I was a virtual rancher of clickable cattle on Facebook. . . .

Facebook's IPO hadn't yet taken place, and its service was still fun to use—although it was littered with requests and demands from social games, like FarmVille and Pet Society.

I'd had enough of it—the click-farming games, for one, but also Facebook itself. Already in 2010, it felt like a malicious attention market where people treated friends as latent resources to be optimized. Compulsion rather than choice devoured people's time. Apps like FarmVille sold relief for the artificial inconveniences they themselves had imposed.

In response, I made a satirical social game called Cow Clicker. Players clicked a cute cow, which mooed and scored a "click."

In creating Cow Clicker, Bogost set out to lampoon FarmVille and the culture that developed around it. In the end, his game triggered the same social forces that had enabled the success of FarmVille. Users got hooked. Bogost got concerned and decided to end the game in an event he called "cowpocalypse." He returned to real life and apparently did not give much thought to Cow Clicker until the news broke about Cambridge Analytica, when he realized that his app had harvested user and friends data, all of which was still on a hard drive in his office. He had forgotten all about it. It seems unlikely that he is alone in that respect.

Bogost's article placed a spotlight on an aspect of Facebook's data security problem that had not received enough public attention to that point: once data leaves Facebook, it is no longer possible to retrieve it, and it may well live on. There is no way now for Facebook to police third parties who harvested data between 2010 and 2014. Someone, somewhere has the data. Maybe not all of it, but most of it. Why wouldn't they? It still has economic value. Data sets may have been sold or given away. Some may have been destroyed. Nobody knows what happened to all that private user data. And no matter how old a data set may be, if you use it inside Facebook again, it will gain benefit from all the data Facebook has acquired since the day the data set was harvested.

The following day, a second whistle-blower emerged from Cambridge Analytica. Unlike Christopher Wylie, who had been at Cambridge Analytica from the start but left prior to the 2016 election, Brittany Kaiser was a senior executive who worked on both Brexit and the US presidential election. She was a political liberal who had joined the 2008 Obama presidential campaign under Facebook cofounder Chris Hughes and who voted for Bernie Sanders in the 2016 Democratic primary. In becoming a whistle-blower, Kaiser said she wanted to stop telling lies. As quoted in *The Guardian*, Kaiser shared her motivation:

"Why should we make excuses for these people? Why? I'm so tired of making excuses for old white men. Fucking hell."

She says she believes that Silicon Valley has much to answer for. "There's a much wider story that I think needs to be told about how people can protect themselves, and their own data."

One of the most significant disclosures in the first interview with Kaiser was about how effective Hughes had been at getting Facebook to implement changes that reduced the workload for the Obama campaign. On behalf of Obama's reelection campaign, Hughes created an application that harvested friends data. The app differed from Kogan's in that it was honest about its true mission, and the mission itself— encouraging people to vote—was admirable, where voter suppression is not, but data harvesting was just as wrong when the Obama campaign did it as it was when Kogan did it four years later.

Kaiser was working at SCL Group when Alexander Nix created Cambridge Analytica as an affiliate. Kaiser's experience in the Obama campaign appealed to Nix, who made a case that the next big market opportunity would be to help Republicans catch up to the Democrats in data analytics. Kaiser transferred into Cambridge Analytica and went to work bringing in clients. Her early clients were in Africa, but in 2015 she and Nix shifted their focus to the United States in anticipation of the presidential election cycle. Kaiser asserted that Nix was not a political ideologue—unlike his patrons Robert Mercer and Steve Bannon—and hoped to create a "famous advertising company in the US market." As quoted in *The Guardian*:

"Corporations like Google, Facebook, Amazon, all of these large companies, are making tens or hundreds of billions of dollars off of monetising people's data," Kaiser says. "I've been telling companies and governments for years that data is probably your most valuable asset. Individuals should be able to monetise their own data—that's their own human value—not to be exploited."

In her *Guardian* interview, Kaiser contradicted Cambridge Analytica's repeated assertions that it had not worked on the Leave campaign during Brexit. Kaiser said that two different organizations affiliated with Leave had entered into data-sharing relationships with Cambridge Analytica. No money had changed hands, she said, but there had been an exchange of value. *The Guardian* explained that such an exchange may have violated UK election law.

The Cambridge Analytica story was growing into a tsunami. Notwithstanding Brexit, the UK government still knew how to conduct an investigation. In all probability, it would not be so easy for Facebook to deflect as the US Congress had been at the hearings in October and November. Facebook was already struggling to manage all the bad news. The threat from the UK would make that much harder.

11

Days of Reckoning

*Facebook's Cambridge Analytica scandal has everything:
peculiar billionaires, a once-adored startup turned
monolith, a political mercenary who resembles a Bond
villain and his shadowy psychographic profiling firm,
an eccentric whistleblower, millions of profiles worth
of leaked Facebook data, Steve Bannon, the Mercers,
and—crucially—Donald Trump, and the results of the
2016 presidential election.* —CHARLIE WARZEL

The United States has many laws and regulations that limit corporate behavior. Passing laws and creating regulations requires enormous effort that happens only in cases of major violations of community standards. Companies are not allowed to dump toxic materials without a permit because society at one time recognized that the consequences of doing so produced unacceptable costs in the form of pollution and damage to public health. Financial institutions are allowed to use customer deposits only in legally authorized ways. Doctors and lawyers are not allowed to share private information provided by clients except in a limited number of strictly specified situations. Companies chafe at regulation, but most accept that rules are a form of friction necessary to protect society. Most executives appreciate the notion that society has the right to limit economic freedom in the interest of the public good, but a tension remains between the freedom of business and the rights of society. Few want limits on their business, and most large businesses retain professional lobbyists and other advocates to protect their interests in the halls of government and in public opinion.

In capitalism, there should be a symbiotic relationship between government and business. Businesses depend on the rule of law, especially property law, to protect their assets. They depend on the government to set the rules and enforce them. In a democracy, the central tension is over which constituencies should have a voice in rule setting. Should it be a private negotiation between businesses and policy makers? Should employees have a voice? How about the communities where businesses operate? Who will protect customers? The aphorism "Let the buyer beware" is helpful, but what happens when the actions of a business affect people who are not customers? These are situations for which laws and regulations exist.

At the federal level, laws are made by Congress and interpreted by courts. To implement laws, agencies in the executive branch create regulations. For example, the Federal Trade Commission regulates consumer protection and aspects of antitrust related to business practices. The Department of Justice addresses antitrust issues related to mergers and large-scale anticompetitive behavior, among many other things. The Department of Labor protects employees. The formulation of new regulations usually involves input from a wide range of constituencies, including the affected businesses. In many cases, businesses are successful at narrowing the scope of regulation from the intent of the original legislation. Each executive-branch agency's mission has evolved in response to new issues, and periodically the country reframes its approach to regulation. The fifty states also play a role in setting the rules that govern business. California conspicuously leads the country in environmental regulations. As the fifth-largest economy in the world, California has unique leverage to enforce emissions standards and other regulatory priorities. Many states have implemented laws and regulations to fill perceived holes at the federal level.

Relative to internet platforms, one conspicuous hole in federal laws and regulations relates to privacy. There is no federal right to privacy in the United States. For most of the country's history, that did not seem like a problem. The Fourth Amendment, which provides protection against unlawful search and seizure, is the closest thing to privacy protection offered in the Constitution. As a result of the privacy gap, many states have implemented privacy-protection laws and regulations to safeguard their citizens.

The current era of federal deregulation has lasted long enough that few business leaders have experience with any other philosophy. Few imagine that regulations can play a constructive role in society by balancing the interests of the masses with those of the rich and powerful. Over time, some regulations become obsolete, but the notion that government "is the problem" overlooks the fundamental importance of sound rules and impartial enforcement to the success of capitalism.

Today, there are few rules and regulations that limit the business activities of internet platforms. As the newest generation in an industry with a long track record of good corporate behavior and products that made life better for customers, internet platforms inherited the benefits of fifty years of trust and goodwill. Today's platforms emerged at a time when economic philosophy in the United States had embraced deregulation as foundational. The phrase "job-killing regulations" had developed superpowers in the political sphere, chilling debate and leading many to forget why regulations exist in the first place. No government or agency creates a regulation with the goal of killing jobs. They do it to protect employees, customers, the environment, or society in general. With an industry like tech, where corporate behavior had been relatively benign for generations, few policy makers imagined the possibility of a threat. They focused on industries with a history of bad behavior and rule breaking. They focused on harms they could see.

Customers expect products and services to provide good value for their money, but the public generally does not look to business for moral leadership. They expect that businesses will compete for profits, taking advantage of whatever tools are available. When harmed by corporate actions, people get angry, but most feel powerless in the face of wealth and power. They feel that different rules apply to the powerful, which is demonstrably true, but there are limits to the harm a business can do without punishment. Even the rich and powerful face legal or regulatory action if they go too far. By the spring of 2018, policy makers and the public were engaged in a lively debate about the internet platforms. Should there be limits on Facebook, Google, and others? Had they gone too far?

If the data set that Cambridge Analytica misappropriated from Facebook had not played a role in the 2016 presidential election, policy makers and the public might have dismissed the story of misappropriated user data as "businesses being businesses." If Facebook employees had not worked with Cambridge Analytica inside the Trump campaign only months after the data-misappropriation scandal first broke in December 2015, Facebook might have had a viable alibi. As things worked out, though, the whole world caught a brief glimpse of an aspect of Facebook that the company had taken pains to hide.

The news hit hard because most users love Facebook. We have come to depend on it. Facebook is not just a tech platform; it has taken on a central role in our lives. As consumers, we crave convenience. We crave connection. We crave free. Facebook offers all three in a persuasive package that offers enough surprise and delight to cause us to visit daily, if not more often. It enhances birthdays, provides access to a wide range of content, and is always available. It enables activists to organize events. Even ads on Facebook can be useful. Cambridge Analytica filled in an unwritten portion of the Facebook story related to the true cost. Convenience and connection on Facebook may not have a sticker price, but they are not free. The downstream costs are substantial but not apparent until something goes wrong, which it has done with alarming frequency since Facebook's founding in 2004. Careless sharing of personal data is terrible, but the story underscored a bigger problem: user data is feeding artificial intelligences whose objective is to manipulate the attention and behavior of users without their knowledge or approval. Facebook's policy of allowing third-party app vendors to harvest friends lists, its tolerance of hate speech, its willingness to align with authoritarians, and its attempts to cover up its role in the Russian election interference are all symptoms of a business that prioritized growth metrics over all other factors.

Was this not just business as usual? Even if the country does not approve of Facebook's choices, do the offenses rise to a level that requires regulatory intervention? Would regulation cause more harm than good? In debating these questions, countries are coming to terms with their disappointment in internet platforms. The fact that we never expected to face these questions makes them especially challenging, but to do nothing would be to turn over stewardship of democracy, public health, privacy, and innovation to a company, and one with a terrible track record.

Facebook has done a brilliant job of converging the virtual world with the real world, but to do so, it has had to rearrange key elements of the social fabric. It has transformed the way people see the world by enabling more than two billion users to have their own reality. It has changed the nature of community by allowing people to sort themselves into like-minded groups where they never have to engage with other viewpoints. It has altered relationships by promoting digital interaction as an alternative to real life. It has manipulated users' attention to increase their level of engagement. It has enabled bad actors to manipulate users and do harm to innocent people on a giant scale. It has provided a platform to bad actors to undermine democracy. Thanks to Cambridge Analytica, users finally had a sense of how much Facebook knew about them and what they did with that knowledge. Users did not like what they learned. Facebook treated private user data as a pawn to be traded for its own advantage.

It is no exaggeration to say that Facebook is one of the most influential businesses in history. Facebook's failures have a profound impact. Hate speech can have fatal consequences, as has been the case in Myanmar and Sri Lanka. Election interference can undermine democracy and change history, as has been the case in the United States and possibly the United Kingdom. When such things happen offline, we send in the police. When they happen online, what is the correct response? How long can we afford to trust Facebook to regulate itself?

Journalists continued to expose examples of Facebook abusing its users' trust. Another disturbing story revealed that users of Facebook on Android discovered their phone data—calls, texts, and other metadata—had been downloaded by Facebook. Presumably Facebook just added the Android data to its oceans of user data. Users had no idea it was happening. While the insecurity of Android has been common knowledge in the industry, it has not prevented the operating system from dominating the cell phone market, with a global share in excess of 80 percent. Like the Cambridge Analytica story, the Facebook/Android news made a security threat real for millions of users.

Now that reporters and users were looking for it, they found examples of bad behavior every day. A particularly ugly example emerged on March 29 in a story from *BuzzFeed*. It described an internal Facebook memo written in January 2016 by Vice President of Advertising Andrew Bosworth, entitled "The Ugly." Written the day after a Facebook Live video captured the shooting death of a man in Chicago, the memo justified Facebook's relentless pursuit of growth in sinister terms.

"We connect people. Period. That's why all the work we do in growth is justified. All the questionable contact importing practices. All the subtle language that helps people stay searchable by friends. All of the work we do to bring more communication in. The work we will likely have to do in China some day. All of it," VP Andrew "Boz" Bosworth wrote.

"So we connect more people," he wrote in another section of the memo. "That can be bad if they make it negative. Maybe it costs someone a life by exposing someone to bullies.

"Maybe someone dies in a terrorist attack coordinated on our tools."

When the article came out, Boz attempted damage control:

I don't agree with the post today and I didn't agree with it even when I wrote it. The purpose of this post, like many others I have written internally, was to bring to the surface issues I felt deserved more discussion with the broader company. Having a debate around hard topics like these is a critical part of our process and to do that effectively we have to be able to consider even bad ideas, if only to eliminate them. To see this post in isolation is rough because it makes it appear as a stance that I hold or that the company holds when neither is the case. I care deeply about how our product affects people and I take very personally the responsibility I have to make that impact positive.

The memo took my breath away. What was he thinking? How could anyone say something like that? What kind of company thinks such language is acceptable? Boz is one of the keepers of the Facebook culture. He is a thought leader inside the company who is known to say provocative things. When he writes a memo, every recipient reads it right away. More important, they take it seriously. Facebook may be much more than Boz's memo and tweet, but the memo and tweet reflect the Facebook culture. The message was clear: the culture of Facebook revolves around a handful of metrics, things like daily users, time on site, revenues, profits. Anything that is not explicitly on the list is definitively off the list. No one at the company allowed him- or herself to be distracted by downstream consequences. They have a hard time imagining why downstream consequences should be Facebook's problem. Boz's memo and tweet were a coup de grace to my idealism about Facebook. The only silver lining was that someone had leaked the memo to the press. If it was an employee, that might signal internal recognition that some of Facebook's internal policies were harmful to society and that management's refusal to cooperate with authorities in the face of clear evidence justified whistle-blowing. Even if the whistle-blower turned out to be a Facebook alum, that was still good news. I have been told the memo came from an alum, but I have been unable to confirm that.

Why don't Facebook employees blow whistles on the company's bad behavior? Why don't users abandon the platform in protest? I cannot explain the behavior of employees, but I understand users. They crave convenience and utility. They struggle to imagine that they would ever be victims of manipulation, data security breaches, or election interference, much less what those things would mean to them. They don't want to believe that the screens they give their children might be causing permanent psychological harm. Elected officials like the campaign technology and contributions they get from Silicon Valley. They like that tech is popular with voters. Confident that they would never need it, policy makers have not developed the expertise necessary to regulate technology. Intelligence agencies do not appear to have anticipated the possibility that the country's enemies might weaponize internet platforms to harm the United States. As a result, no one was prepared for election interference, hate speech, and the consequences of addiction.

Major technology companies have exploited both users' trust and the persuasive technology in their platforms to minimize political fallout and protect their business models. Until Cambridge Analytica, it worked.

As new stories emerged almost daily that reinforced the narrative that Facebook had failed at self-regulation, Zuck rejected calls for heads to roll, saying that he was the best person to run the company and was responsible for what had gone wrong. *The Washington Post* reported on a Facebook admission that "malicious actors" had exploited the search tools on its platform to "discover the identities and collect information on most of its 2 billion users worldwide." Then we learned that a change Facebook made in the way that third-party apps interact with its system had inadvertently locked people out of Tinder, a dating app that uses Facebook for authentication and other personal data. (This story soon seemed like foreshadowing when, a month later, Facebook announced a dating application.) On April 4, Facebook announced the first revision to its terms of service since 2015. The bulk of the changes related to disclosures with respect to privacy and the handling of user data. Small steps, but progress. Public pressure was working.

One particularly awkward story that week revealed that Facebook had been deleting Zuck's Messenger messages from the inboxes of recipients, a feature not available to users. Facebook initially claimed it made the change for security purposes, but that was patently unbelievable, so the next day it announced plans to extend the "unsend" feature to users. (A limited "unsend" capability shipped in February 2019.) The company then announced greater transparency for political ads, following the model of the Honest Ads Act introduced by Senators Mark Warner, Amy Klobuchar, and John McCain but extending it to include ads supporting issues as well as candidates. Uniquely among Facebook's recent changes, this one stood out for being right on substance, as well as on appearances. While false flag ads had played a relatively small role in the Russian interference in 2016, that role had been essential to attracting American voters into Russian-organized Facebook Groups, which in turn had been a major tool in the interference. For the new policy to work, Facebook will have to change its approach to enforcement.

With pressure building, journalists and technologists seemed nearly unanimous in their view that members of Congress did not understand technology well enough to regulate it. Many suggested that regulating technology was a fool's errand in any case, as it would distort the marketplace in a way that protected the largest incumbents at the expense of smaller players and startups. That concern was valid but argued for care in formulating regulations, not for laissez-faire. Congress's lack of experience in regulating tech does not absolve it of its responsibility to protect Americans from the failures of the marketplace. In reality, tech is less complicated than health care, banking, and nuclear power. It just changes faster. Every time a new industry requires regulation, Congress must get up to speed. The challenge would be to develop the necessary skills quickly. This would not be the first time Congress faced that challenge.

Public pressure produced more concessions from Facebook, which announced additional policy and product changes in an attempt to appear cooperative and preempt regulatory action. As usual, the announcements featured sleight of hand. First, Facebook banned data brokers. While this sounded like a move that might prevent future Cambridge Analyticas, what it actually did was move Facebook closer to a data monopoly on its platform. Advertisers acquire data from brokers in order to improve ad targeting. By banning data brokers, Facebook forced advertisers to depend entirely on Facebook's own data. As the month of March ended, Facebook posted on its blog an update about its effort to prevent future election interference. The post focused on Facebook's plans to stop bad actors from hiding behind false identities, as well as a new initiative to anticipate interference rather than waiting for users to report it. While the announcement included a new program to target disinformation, the change to News Feed in January that reduced the weight of journalistic sources almost certainly made at least one aspect of preventing interference—the piercing of filter bubbles—much harder.

Zuck agreed to testify at two congressional hearings—a joint session of the Senate Judiciary and Commerce committees, the other with the House Committee on Energy and Commerce—but refused to appear before the UK Parliament.

The hearings began on the afternoon of April 10 in the Senate. Zuck arrived in a suit, shook lots of hands, and settled in for five hours of questions. The combined committees have a total of forty-five members. Each senator would have only four minutes, which favored Zuck, who prepared well for the format. If he could be long-winded with each answer, Zuck might be able to limit each senator to only three or four questions. Perhaps more important, the most senior members of the committee went first, and they were not as well prepared as Zuck. Whether by luck or design, Facebook had agreed to appear on the first day after a two-week recess, minimizing the opportunity for staff members to prepare senators. The benefits of that timing to Facebook were immediately obvious. Several senators did not seem to understand how Facebook works. Senator Orrin Hatch asked, "How do you sustain a business model in which users don't pay for your service?" revealing his ignorance about Facebook's advertising business model. Armed with a cheat sheet of diplomatic answers, Zuck patiently ran out the clock on each senator. Senators attempted to grill Zuck, and in the second hour, a couple of senators asked pointed questions. For the most part, Zuck deflected. Zuck also benefited from a lack of coordination among the senators. It seemed that each senator addressed a different issue. Perhaps a dozen different problems emerged, each of which might have justified its own hearing.

Early in the third hour, Zuck caught a lucky break: networks interrupted their coverage to reveal that the FBI had raided the home and office of Donald Trump's personal attorney, Michael Cohen. News outlets pivoted instantly. Hardly anyone saw the rest of the Senate hearing or any of the House Committee on Energy and Commerce hearing the next day. For the five-hour House hearing, Zuck used the same "run out the clock" strategy. Without a big TV audience, it probably didn't matter, but the Democratic House members landed some punches. Several coordinated their questions, some of which caught Zuck unprepared. Representative Frank Pallone of New Jersey pushed Zuck for a yes-or-no answer as to whether Facebook would change its default settings to minimize data collection. Zuck equivocated. He would not answer Pallone directly, presumably because Facebook has no intention of minimizing data collection. For Zuck, the hearing went downhill from there.

Representative Mike Doyle of Pennsylvania focused on Cambridge Analytica and the harvesting of user data. In response to a question from Doyle, Zuck claimed that he first learned that Cambridge Analytica had acquired the data set when *The Guardian* reported the story in December 2015. Doyle gently suggested that perhaps Facebook was not paying attention to such things. He observed, "It seems like you were more concerned with attracting and retaining developers on your platform than you were with ensuring the security of Facebook user data."

Representative Kathy Castor of Florida grilled Zuck on the breadth of the company's data collection, both on and off the platform.

"For all of the benefits that Facebook has provided in building communities and connecting families, I think a devil's bargain has been struck," she said. "And, in the end, Americans do not like to be manipulated. They do not like to be spied on. We don't like it when someone is outside of our home, watching. We don't like it when someone is following us around the neighborhood or, even worse, following our kids or stalking our children.

"Facebook now has evolved to a place where you are tracking everyone. You are collecting data on just about everybody. Yes, we understand the Facebook users that—that proactively sign in, they're in part of the—that platform, but you're following Facebook users even after they log off of that platform and application, and you are collecting personal information on people who do not even have Facebook accounts. Isn't that right?"

According to *Politico*, Castor's questioning left Zuck "looking particularly on edge." She focused on a foundational element of Facebook's business: when it comes to data privacy, user choice is an illusion. The terms of service protect Facebook, not the user. Castor's questions revealed that she recognized Congress has a duty to address digital surveillance to protect consumers.

Representative Ben Ray Luján of New Mexico dug much deeper into the issue of data collection. When Zuck professed not to know how many data points Facebook had on the average user, Luján told him: twenty-nine thousand. Luján also pinned Zuck down on a paradox: Facebook collects data on people who do not use the platform and have no ability to stop Facebook from doing that without themselves joining Facebook.

Representative Joe Kennedy of Massachusetts focused on metadata. His questions reflected concern that users have no idea how much data Facebook collects or that metadata can be used to paint a detailed picture of the user. Kennedy's questions highlighted that users have some control over the content they post on Facebook but not over the metadata created by their activity, which is the fuel for Facebook's advertising business.

The House hearing exposed the extent of Facebook's data collection and its lack of regard for user privacy. For those who watched the hearing, it seemed that Facebook has lots of policies that are designed to protect the company from legal challenges without actually accomplishing anything else. When challenged, Zuck often professed ignorance. He repeatedly promised to have his staff follow up with a response. He apologized for Facebook's lapses. It didn't matter. The verdict in the press—and on Wall Street—was formed in the first hour of the Senate hearing, and it was nearly unanimous: Zuck had exceeded expectations, while Congress fell short. Many commentators cited the hearings as evidence that Congress lacked the sophistication necessary to regulate technology. Based on follow-up meetings with members, I learned that the Democrats in Congress recognized the need for real oversight of internet platforms. To prepare, they needed help from experts without conflicts and asked if our team could help. The age of innocence was over.

12

Success?

*Everybody gets so much information all day long that
they lose their common sense.* —GERTRUDE STEIN

Months of daily revelations about Facebook's business model and choices drove the conversation about the dark side of social media into the public consciousness, but that did not mean users understood why it should matter to them. Facebook had announced product and policy changes in response to public pressure, but the threat to public health, democracy, privacy, and innovation remained. Reforming the behavior of internet platforms remained an uphill battle. To use the ascent of Everest as an analogy, we had hiked to an elevation of 17,598 feet and had only reached Base Camp. The hard part is in front of us.

Internet platforms still enjoyed a massive reservoir of goodwill with policy makers and the public, and they leveraged it where they could. Their challenge was made easier by the wide range of harms. It was hard to keep up. Anecdotes like Cambridge Analytica, Russian election interference, ethnic cleansing in Myanmar, and the rising suicide rate among teens attracted attention, but most users could not understand how products they trusted could possibly have caused so much harm. Understanding how the choices made by internet platforms had caused these things to happen would require more time and effort than most users were willing to commit. Figuring out what to do about it would be even more challenging.

Tech had become so powerful that by pointing it at the weakest links in human psychology, platforms could manipulate their users' attention for massive profit. Whether they acknowledged it or not, the platforms had demonstrated the ability to influence huge groups, including whole countries. Policy makers and the public had to decide whether this kind of business activity violated societal norms. Scientists have the ability to create deadly viruses, but society does not permit them to do so, except in carefully controlled research settings. Financial services businesses have the ability to defraud their customers, but society does not allow that either. What should be the limits, if any, on internet platforms? The next phase of the conversation needed to address the proper role of internet platforms in society. How far should platforms be allowed to go in exploiting human weakness? How much responsibility should they bear for harm that results from their products and their business model? Should platforms bear fiduciary responsibility for the user data they hold? Should there be limits on the exploitation of that data?

The issues go far beyond Facebook. Google's surveillance engine gathers more data on users than any other company. YouTube has become the nexus for recruiting and training extremists. It is home to countless conspiracy theories. It hosts age-inappropriate content targeting little children. Instagram and Snapchat magnify the anxieties of teenagers in ways both small and large. But as the dominant social network, Facebook has benefited most from exploiting the lack of awareness and regulation of online business practices. Its scale magnifies its failures. Not all of the most extreme problems are on Facebook, but the platform has been a magnet for bad actors because of its reach, its unguarded nature, and its cultural indifference to the downstream consequences of its actions.

With 2.4 billion active users, Facebook rivals the world's largest religion in terms of scale, and that number does not include its subsidiary platforms. It has huge influence in every country in which it operates. In some countries, like Myanmar, the internet essentially is Facebook. The persecution of the Rohingya minority continues in that country, and Facebook has been cited by United Nations investigators for enabling "ethnic cleansing." In August 2018, Reuters released a special report that uncovered more than "1,000 examples of posts, comments, and pornographic images attacking the Rohingya and other Muslims" even after the platform had taken steps to prevent hate speech in Myanmar. Later that month, Facebook banned Myanmar military accounts from the site. While the ban is unlikely to prevent hate speech in Myanmar, it is nonetheless historic. It is the first time that I am aware of that Facebook acted against the powerful on behalf of the powerless. Until there is evidence to the contrary, I will assume the ban reflects a short-term desire for a public relations win rather than a major change in Facebook policy.

Despite a huge increase in focus and resources, there is no indication yet that Facebook has been able to stop the hate speech on its platform in Myanmar or any other country. As a result, the Rohingya continue to suffer. In the United States, Facebook has become the most important platform for news and politics. The impact of Facebook on public discourse is unprecedented, thanks to its *Truman Shows*, filter bubbles, and manipulation of attention. No one elected the employees of Facebook, but their actions can have a decisive impact on our democracy. In almost every democracy, Facebook has become an unpredictable force in elections. For authoritarian regimes, Facebook has become a preferred tool for controlling the citizenry. This has been demonstrated in both Cambodia and the Philippines. The government of China is taking the idea to its limit, creating a social media platform based on "social credit," which rewards users for actions approved by the government, while penalizing actions that are not approved. The goal of the Chinese project appears to be behavior modification on a national scale.

When the conversation about the appropriate role of internet platforms began in 2017, Facebook chose to fight rather than listen to its critics. It tried to bluff its way past criticism and failed. Facebook suffered some brand damage but did not alter its course. The pressure grew, and Facebook responded, first with small changes to its business practices, then with larger ones. I have no doubt that by the middle of 2018 Facebook was doing its best to fix many of the problems that have come to light, but only did so in ways that protect its business model and growth. At this writing, it has not made the fundamental changes necessary to prevent future election interference, limit manipulation of its users by third parties, prevent hate speech, or protect users from the consequences of Facebook's willingness to share user data. These issues are now known, if not fully understood. The question is whether policy makers and users will insist on change.

After Zuck's performance at the hearings, Facebook may have concluded that the storm had passed. Confirmation came a couple of weeks after the hearings, when Facebook reported earnings for the first quarter of 2018. It was a blowout. Facebook's key metrics showed big improvement. Revenue for the quarter came in just below $12 billion, half a billion more than analysts had forecast. Profits jumped 63 percent. Monthly active users of 2.2 billion and daily active users of 1.47 billion both rose 13 percent from the prior year. The company finished the quarter with forty-four billion dollars in cash and marketable securities. On the conference call with investors, Zuck made a quick reference to the hearings and Cambridge Analytica, but anyone expecting a mea culpa would have been disappointed. In combination with the reviews of his testimony before Congress, the earnings report restored Zuck to his happy place. Everything about his conference call remarks proclaimed that Facebook had returned to business as usual. Investors could not have been happier. The stock jumped 9 percent the next day.

But bad news continued to surface, including stories about illegal ivory trading on Facebook and fake Facebook pages that scammed Vietnam veterans. As late arrivals to a riot of bad behavior, the new stories had little impact. They were joined by a second, more analytical wave of postmortems from the congressional hearings that took some of the shine off Zuck's performance, at least with policy makers.

The Washington Post published a lengthy article that fact-checked Zuck's testimony. It detailed the way that Zuck repeatedly reframed questions to hide as many of Facebook's unattractive behaviors as possible. For example, when asked questions about users' ability to access the data Facebook has about them, Zuck focused exclusively on content data, which is generally accessible, avoiding wherever possible any reference to the third-party data that drives its advertising business and is generally not viewable or manageable by users. He was extremely effective at this kind of reframing, particularly in the Senate hearing. The public probably missed this story, but the members of Congress charged with oversight did not.

The hearing made it clear that for those of us who believe in democracy and who want to protect public health and privacy, there would be a long journey ahead. Facebook had come through two huge scandals—the Russian interference and Cambridge Analytica—and two sets of congressional hearings, with only a few dings in its reputation to show for it. The business itself was running at full speed, unimpeded by all the criticism. It was going to take a lot more than a scandal and some hearings to persuade Facebook to make the fundamental changes necessary to protect users from harm.

The truth of this insight became obvious the first week of May, at F8, Facebook's annual conference for third-party developers. For Zuck and Facebook, F8 was a triumph. Zuck mentioned Cambridge Analytica, but it occupied roughly the same proportion of his speech as vermouth in a very dry martini at the bar of a private club. Otherwise, the event was a wall-to-wall lovefest. Developers showed zero concern about the recent scandals. Facebook exhibited no remorse. If anything, the company behaved as if surviving the crisis had made it stronger. Perhaps it had.

Facebook announced new product initiatives at F8, including a dating service and Clear History, a tool for seeing and erasing the browsing data the company has accumulated on you. The dating service offered a new wrinkle to the market—a focus on events—and had an immediate impact on the market leader, Match.com, which owns the eponymous site, as well as Tinder. Match's stock dropped 22 percent when the news broke, in no small measure because its sites, as well as those of competitors, had long leveraged Facebook for user authentication and data. As music apps, games, and news publishers before it had learned the hard way, trusting Facebook to be a good partner generally leads to disappointment. Eventually Facebook will undermine the economics of your business.

Clear History may be a great idea, but Facebook's recent behavior also suggested the possibility that the announcement was just another public relations stunt. Facebook deploys trackers—tiny pieces of code dropped into the browser—to follow users around the web. They do this not only from their various platforms but also from the Facebook Connect log-in tool and the millions of Like buttons that are sprinkled all across the World Wide Web. For users who leverage Facebook credentials to log in to other sites and touch Like buttons where they go, this translates into a massive trove of browsing history and metadata that can be used to construct a high-resolution image of the user that proved to be the key to making Facebook's advertising valuable. In offering users a Clear History app, Facebook is signaling one of two things: either it is finally taking user concerns about privacy to heart, or it no longer needs to maintain browser histories in order to sell ads.

The former would be amazing, but I do not think we can rule out the latter. After an eighteen-month wait, Facebook began to roll out Clear History in August 2019. Public pressure helped to bring it about.

In a conversation with Representative Joe Kennedy days after the House hearing, Zuck indicated that Clear History would apply to metadata as well as links, which would represent a huge departure from recent practice and a genuine benefit to users. Skeptics point to a more ominous explanation. Facebook has been using its massive store of user data to train the behavioral-targeting engine of its artificial intelligence. In the early phases of training, the engine needs every piece of data Facebook can find, but eventually the training reaches a level where the engine can anticipate user behavior. Perhaps you have heard anecdotes about people saying a brand name out loud and then seeing an ad for that brand on Facebook. The users assume Facebook must be using their device's microphone to eavesdrop on conversations. That is not practical today. A more likely explanation is that the behavioral-prediction engine has made a good forecast about a user desire, and the brand in question happens to be a Facebook advertiser. It is deeply creepy. It will get creepier as the technology improves. Once the behavioral-prediction engine can forecast consistently, it will no longer need the same amount of data that was required to create it. A smaller flow of data, much of it metadata, will get the job done. If that is where we are, letting users clear their browsing history on Facebook would provide the illusion of privacy, without changing Facebook's business or protecting users.

From our first days together, Tristan convinced me that among the many dark sides of social media, artificial intelligence engines might pose the greatest threat to society. He argued that the AIs of companies like Facebook and Google have insurmountable advantages: infinite resources and scalability; an exceptionally detailed profile of more than two billion users (including, in Facebook's case, a deep understanding of emotional triggers); a complete picture of each user's location, relationships, and activities; and an economic incentive to manipulate user attention without regard to consequences. Effectively the AI has a high bandwidth connection directly into the cerebral cortex of more than two billion humans who have no idea what they are up against. When the AI behavioral-prediction engines of Facebook and Google reach maturity, they may be able to abandon the endless accumulation of content data and some forms of metadata—addressing a meaningful subset of the privacy concerns that have been raised in the press and in congressional hearings—without actually improving users' privacy. That would be a terrible outcome. Under intense political pressure to "do something" about privacy, policy makers might rush to make changes that would create only the illusion of benefit. Users need privacy protection that starts with prior, informed consent for any use of their data. Without it, all users will be vulnerable to bad actors who can use the power of the internet to smear reputations and cause irreparable harm.

Judy Estrin, technologist, entrepreneur and networking pioneer, believes that we need to take a good look at our relationship with technology. She points out that the combination of free-market capitalism, network effects, and platform monopolies, plus trust in tech by users and policy makers, has left us at the mercy of "technological authoritarianism." The unelected leaders of the largest technology platforms—but especially Facebook and Google—are eroding the foundations of liberal democracy around the world, and yet we have entrusted them with the information security of our 2020 election. They are undermining public health, redefining the limits of personal privacy, and restructuring the global economy, all without giving those affected a voice. Everyone, but especially technology optimists, should investigate the degree to which the interests of the internet giants may conflict with those of the public.

Policy makers must expect intense opposition from the industry. Some will be tempted to bow to lobbyists. The industry is entitled to a voice in policy, but current law gives them disproportionate power in the regulatory process. They will try to narrow the focus and direct it in ways that minimize the impact on their business. That is their right, but they are not guaranteed the last word. The public can and should ensure that its voice is decisive. Public pressure is already having an effect, and more pressure can have a greater effect. Users can also influence internet platforms by changing their online behavior.

I have come to appreciate that in rapidly changing industries regulation is a blunt instrument. Policy makers understand this, which is one reason why they are reluctant to regulate an industry like tech. Again, the goal is to improve behavior by changing incentives. In a perfect world, the threat of regulation would be enough to accomplish that. When threats do not work, policy makers usually start with the easiest, least painful regulations. If they fail, each new round of regulation will become progressively more onerous. For this reason, the target industry is usually smart to embrace the process early, cooperate, and try to satisfy the political needs of policy makers before the price gets too high. For Facebook and Google, the first "offer" was Europe's General Data Protection Regulation (GDPR). Had they embraced it fully, their political and reputational problems in Europe would have been reduced dramatically, if not eliminated altogether. For reasons I cannot understand, both companies have done the bare minimum to comply with the letter of the regulation, while blatantly violating the spirit of it.

In 2018, I received an email from Representative Zoe Lofgren, a member of Congress from the Bay Area. A strong supporter of the platforms during their honeymoon period, Representative Lofgren told me she was working on an internet privacy bill of rights. Her approach was straightforward:

> In a free society, people must have a right to privacy. To promote that freedom and privacy, we declare that you ***own your data***.
>
> Accordingly, you have a right in a clear and transparent manner to:

(1) consent or opt in when personal information is collected or shared with a third party and to limit the use of personal information not necessary to provide the requested service.

(2) obtain, correct or delete personal data held by a company.

(3) be notified immediately when a security breach is discovered.

(4) be able to move data in useable, machine-readable format.

I loved the simplicity and directness of this data bill of rights. Similar in objectives and approach to a bill introduced by Senators Klobuchar and Kennedy, as well as to Europe's GDPR, Representative Lofgren's proposal could be a valuable first step in an effort to regulate the platforms. Getting any data privacy bill of rights through Congress in the current environment would be an extreme challenge. Passing a bill that would actually restore ownership and control to users would be even harder. Congress would face intense lobbying from the platforms, who, no matter their public posture, would do their best to weaken any bill that limited their freedom to act. The long odds are not an argument against trying to pass a bill, but they do create an incentive not to waste energy on bad legislation. At this writing, Facebook and Google are doing precisely that, pushing Congress to implement underpowered privacy regulations.

In my response, I congratulated Representative Lofgren and provided some feedback:

Relative to #4, I think it is important to specifically secure the rights to all metadata—the full social graph with relationships and actions, in addition to names—in a portable fashion. Without that, startups are going to be disadvantaged and innovation will be limited to some degree.

Whatever bill of rights we create has to focus as much on how data is used as on how it is collected.

This last point reflected my concern that Facebook, Google, and others have enabled reams of private user data to escape their networks. There is no way to reclaim that data or even to know where it has gone. The best we can do is protect users from unforeseen and inappropriate uses of the data. Any internet bill of rights that does not address uses of data may have little if any practical value to users.

In addition, I asked Representative Lofgren to embrace the view that no matter how well implemented, privacy regulations would address only a subset of the problems created by platforms. For example, I hypothesized that the lack of competition in core social media categories narrowed the scope and slowed the pace of innovation. There is no alternative to Facebook or YouTube. If you don't like their business practices, you are stuck. The platforms have been able to acquire promising startups in adjacent categories—as Facebook had done with Instagram and WhatsApp—converting potential competitors into extensions of their monopoly. In addition, Facebook and Google have gotten footholds in many promising new categories—ranging from virtual reality to AI to self-driving cars—in their pre-market stages. Engagement by the platforms has validated new categories, but in all probability, it has also distorted them, changing the incentives for market participants. It is hard to imagine that Facebook's purchase of and commitment to the Oculus virtual-reality platform did not discourage investment in alternative hardware platforms. What venture capitalist would want to compete with Facebook in a category that might require an investment of hundreds of millions of dollars? It is also hard to believe that early-stage projects inside giant platforms operate with the same sense of urgency as startups, which suggests that the pace of innovation might suffer.

From my seat in Silicon Valley, I can see that the success of Google, Amazon, and Facebook has distorted the behavior of entrepreneurs and investors. Entrepreneurs have a simple choice: stay away from the giants or create businesses designed to be sold to them. The result has been a flood of startups that do things your parents used to do for you, including all manner of transportation and delivery services, as well as services that do things like clean up after your dog. Many new products seem to be targeted at billionaires, as was the case with Juicero's seven-hundred-dollar juicer. In an environment that is already challenging for entrepreneurs and startups, it will be important for policy makers not to make the situation worse.

Regulations like the GDPR or Representative Lofgren's privacy bill of rights impose a cost of compliance with a fairly high minimum that applies to all affected companies, irrespective of size. Without modification, the costs may be disproportionately burdensome on startups, further enhancing the competitive advantages of the largest companies, but there are ways to compensate for that without undermining a very important new regulation. My feedback to Representative Lofgren recommended modifying her article #4 to allow portability of the entire social graph—the entire friend network—as a way to promote competition from startups. If you want to compete with Facebook today, you have to solve two huge problems: finding users and then persuading them to invest in your platform to reproduce some of what they already have on Facebook. Portability of the social graph—including friends—would reduce the scope of the second problem to manageable levels, even when you factor in the need for permission from every friend. But graph portability was just the first step. I also advocated antitrust measures.

In my message to Representative Lofgren, I proposed the adoption of a classic model of antitrust as the least harmful, most pro-growth form of intervention she could advocate. I had just written an op-ed for the *Financial Times* on the subject of the 1956 consent decree with

AT&T, which ended that company's first antitrust case, but it had not yet been published, so I gave Representative Lofgren a preview. The decree had two key elements: AT&T agreed to limit itself to its existing regulated markets, which meant the landline telephone business, and it agreed to license its patent portfolio at no cost. By limiting itself to regulated markets, AT&T would not enter the nascent computer industry, leaving that to IBM and others. This was a very big deal and was consistent with historical practice. AT&T owed its own existence to a prohibition on telegraph companies entering telephony. Allowing the computer industry to develop as its own category proved to be good policy in every possible way.

Compulsory licensing of the AT&T patent portfolio turned out to be even more important. AT&T's Bell Labs did huge amounts of research that led to a wide range of fundamental patents. Included among them was the transistor. By making the transistor available for license, the 1956 consent decree gave birth to Silicon Valley. All of it. Semiconductors. Computers. Software. Video games. The internet. Smartphones. Is there any way that the US economy would have been better off allowing AT&T to exploit the transistor on its own timeline? Does anyone think there is a chance AT&T would have done as good a job with that invention as the thousands of startups it spawned in Silicon Valley? Here's the clincher: the 1956 consent decree did not prevent AT&T from being amazingly successful, so successful that it precipitated a second antitrust case. The company was ultimately broken up in 1984, a change that unleashed another tsunami of growth. Postbreakup, every component of the old monopoly flourished. Investors prospered. And two new industries—cellular telephony and broadband data communications—came to market far sooner than would otherwise have been the case.

Applying the logic of the 1956 AT&T consent decree to Google, Amazon, and Facebook would set limits to their market opportunity, creating room for new entrants. That might or might not require the divestiture of noncore operations. There is nothing in the patent port-folios of the platform giants that rivals the transistor, but there is no doubt in my mind that the giants use patents as a defensive moat to keep competitors at bay. Opening up those portfolios would almost certainly unleash tremendous innovation, as there are thousands of en-trepreneurs who might jump at an opportunity to build on the patents.

In my message to Representative Lofgren, I forgot to include some-thing important. Harvard professor Jonathan Zittrain had written an op-ed in *The New York Times* that recommended extending to data-intensive companies the fiduciary rule that applies to professions that hold sensitive data about clients. As fiduciaries, doctors and lawyers must always place the needs of the client first, safeguarding privacy. If doctors and lawyers were held to the same standard as internet plat-forms, they would be able to sell access to your private information to anyone willing to pay. Extending the fiduciary rule to companies that hold consumer data—companies like Equifax and Acxiom, as well as internet platforms—would have two benefits. First, it would create a compelling incentive for companies to prioritize data privacy and secu-rity. Second, it would enable consumers (and businesses) harmed by data holders to have a legal remedy that cannot be unilaterally elimi-nated by companies in their terms of service. Today, the standard prac-tice is to force users who feel they have been harmed to go into arbitration, a process that has historically favored companies over their customers. If consumers always had the option of litigation, companies would be less likely to act carelessly. The fiduciary rule had another benefit: simplicity. It would not require a new bureaucracy or even a complicated piece of legislation.

In the last week of April 2018, I returned to Washington for several days of meetings and events. I met with a number of members of Con-gress, including Representative Ro Khanna, a friend from Silicon Val-ley who had joined forces with Zoe Lofgren on the data privacy bill of rights. Ro was really interested in the fiduciary rule, and we discussed how it might be paired with a bill of rights to make the former more effective in protecting consumers and encouraging competition.

Later that day, I had an impromptu conversation with Nancy Pelosi, who wanted to be certain that members of her caucus had access to insight from sources besides industry lobbyists. We discussed the work I had been doing on the data privacy bill of rights with Representatives Lofgren and Khanna. She let me know that the next step would be to engage with members of the Energy and Commerce Committee, as any such legislation would need to begin in that committee. She agreed to organize a meeting with key members of the committee so that we could share our ideas not only on data privacy but also on the fiduciary rule, antitrust, election security, and public health. In response to Pelosi's concern about counterbalancing the influence of lobbyists, I offered to organize a curriculum with coaches, much as we had done with the staff of the House Permanent Select Committee on Intelligence prior to the November 1 hearing. The goal was to prepare the Democrats in the House for the future, for the day when they would be back in power. Pelosi's goal was to ensure her party would be well prepared to exercise its oversight responsibility.

Washington is just one venue where change can occur. Another is the states. State attorneys general have subpoena power and the ability to bring legal cases where they find violations of consumer rights or other malfeasance. We first met with the attorney general of New York and his staff in 2017, and we expanded our effort to include the AG offices in Massachusetts, California, Washington, and Maryland. A few of these offices dug into Facebook's representations to users with respect to data privacy. Only time will tell if a case emerges. In addition, we joined with our partners at Common Sense Media to work with members of the California legislature on bills to protect user privacy and regulate bots. California has a long history of leadership in such matters and began to extend that leadership to the regulation of internet platforms in June 2018 with the nation's strongest digital privacy law.

On the afternoon of July 25, Facebook reported its second quarter earnings. The big news was that user count was flat in North America and declined in Europe, the two most profitable markets for Facebook. Usage of the Facebook.com website declined precipitously in North America, more than offsetting the growth of the mobile app. The stock declined about 20 percent the following day, losing $120 billion in market value, the largest one-day loss of value in history. While investors did not show any concern about Facebook's business practices, they punished the stock for what may be nothing more than market saturation, which does not have to be a bad thing for a monopolist. To me, the real question is whether a lower stock price might cause employees to reconsider the company's strategy and their role in it. Might it produce whistle-blowers?

At the end of July, Facebook announced that it had shut down thirty-two false pages and profiles that employed the same tactics that the Russians used to interfere in the 2016 election. The pages and profiles worked in a coordinated way across both Facebook and Instagram. Two hundred ninety thousand Facebook users had interacted with the pages, one of which had ties to the Russian Internet Research Agency. In early August, Apple, Facebook, and YouTube removed content and pages associated with Alex Jones and Infowars for violations of unspecified rules regarding hate speech; Twitter did not initially follow suit. The action was an abrupt reversal for Facebook and YouTube. Facebook, in particular, had long argued that it was a neutral platform and did not want to be in the business of deciding what content was appropriate on its site, leaving such decisions to amorphous "community standards." Other than a few obvious categories like child pornography, Facebook has avoided judging the content on its site. This policy had the effect of enabling high-engagement content like conspiracy theories and disinformation to flourish, with significant benefits to profitability. A hypothesis emerged that Apple's announcement—which had preceded the other two by a matter of hours—had given political air cover to Facebook. Twitter attempted to defend its initial inaction on free speech grounds, which unleashed a tsunami of criticism. Twitter responded by suspending Jones for a week, and later banned him and Infowars after Jones was streamed harassing and insulting a CNN reporter who was covering the appearance of Twitter's CEO, Jack Dorsey, at a Congressional hearing. One might reasonably conclude that the internet platforms do not want to damage their business models by banning conspiracy theories, disinformation, and hate speech, unless it becomes politically impossible not to.

Later in August, *The New York Times* reported that Facebook deleted more than 652 fake accounts and pages with links to Iran that were operating a coordinated campaign to influence US politics. According to CNN, the campaign included 254 Facebook pages and 116 Instagram accounts that had more than one million followers. Unfortunately, Facebook did not discover the accounts and pages, despite the investment it has made in election security; a cybersecurity firm named FireEye made the initial discovery. At Facebook's current scale, I believe it's not enough to rely on third parties to identify bad actors—too much harm gets done before the process can stop the problem.

Shortly after the August deletions, the website Motherboard published an investigative story about the way that Facebook moderates content. Facebook gave Motherboard access to the moderation team, which Motherboard supplemented with leaked documents, possibly the first to come from active Facebook employees. According to the story's authors, Jason Koebler and Joseph Cox, "Zuckerberg has said that he wants Facebook to be one global community, a radical ideal given the vast diversity of communities and cultural mores around the globe. Facebook believes highly-nuanced content moderation can resolve this tension, but it's an unfathomably complex logistical problem that has no obvious solution, that fundamentally threatens Facebook's business, and that has largely shifted the role of free speech arbitration from governments to a private platform."

Facebook told Motherboard that its AI tools detect almost all of the spam it removes from the site, along with 99.5 percent of terrorist-related content removals, 98.5 percent of fake account removals, 96 percent of nudity and sexual content removals, 86 percent of graphic violence removals, and 38 percent of hate speech removals. Those numbers sound impressive, but require context. First, these numbers merely illustrate AI's contribution to the removal process. We still do not know how much inappropriate content escapes Facebook's notice. With respect to AI's impact, a success rate of 99.5 percent will still allow five million inappropriate posts per billion. For context, there were nearly five billion posts a day on Facebook . . . in 2013. AI's track record on hate speech—accounting for 38 percent of removals—is not helpful at all. Human moderators struggle to find inappropriate content missed by AI, but they are forced to operate within the constraints of Facebook's twin goals of maximum permissiveness and generalized solutions to all problems. The rule book for moderators is long and detailed, but also filled with conflicts and ambiguity. Moderators burn out very quickly.

The Motherboard reporters concluded that Facebook is committed to getting moderation right, but will not succeed in doing so on their terms. The academics quoted in the article argued that moderation is not possible at Facebook's scale with Facebook's approach. The network is too complex. Facebook has not yet accepted this reality. They continue to believe that there is a software solution to the problem and that it can be successful without changing their business model or growth targets. Facebook is certainly entitled to that point of view, but policy makers and users should be skeptical.

Zuck and Sheryl must know that Facebook's brand has suffered from the onslaught of bad news. To get a sense of the impact, I asked Erin McKean, founder of Wordnik and former editor of the *Oxford*

Dictionary of American English, to study changes in the nouns and adjectives mostly frequently associated with each of the largest tech companies: Apple, Google, Amazon, Facebook, and Microsoft, plus Twitter. Prior to the 2016 election, the tech leaders enjoyed pristine reputations, with no pejorative word associations. For Google, Amazon, Apple, and Microsoft, that is still true. For Facebook, things have changed dramatically. The word "scandal" now ranks in the top 50 nouns associated with Facebook. "Breach" and "investigation" are in the top 250 nouns. With adjectives the situation is even worse. Alone among the five tech leaders, Facebook had one pejorative adjective in its top 100 in 2015–2016: "controversial." In 2017 and 2018, the adjective "fake" ranked in the top 10 for Facebook, followed by "Russian," "alleged," "critical," "Russian-linked," "false," "leaked," and "racist," all of which ranked in the top 100 adjectives. Apple, Google, Amazon, and Microsoft do not have a single pejorative noun or adjective on their lists. Twitter has two nouns on its list that may or may not imply brand issues: "Trump" and "bots." The study was conducted using the News on the Web (NOW) corpus at Brigham Young University. The top 10 US sources in the corpus, ranked by number of words, are *Huffington Post*, NPR, CNN, *The Atlantic*, *TIME*, *Los Angeles Times*, *Wall Street Journal*, *Slate*, *USA Today*, and ABC News.

Despite all the political fallout, Facebook continues to go about its business. In early August, the *Wall Street Journal* reported that Facebook had asked major banks "to share detailed financial information about their customers, including card transactions and checking-account balances, as part of an effort to offer new services to users." Among the banks were JP Morgan Chase, Wells Fargo, and U.S. Bancorp. At least one large bank "pulled away from the talks due to privacy concerns." Facebook spokesperson Elisabeth Diana disputed the report, saying, "A recent *Wall Street Journal* story implies incorrectly that we are actively asking financial services companies for financial transaction data—this is not true." She asserted that Facebook's goal is to integrate chat bots from Messenger with bank services, so that customers can look up their balances and the like.

Also in early August, Facebook announced a new organization called Facebook Connectivity, an umbrella for its many efforts to bring onto its service the remaining 4 billion unconnected humans on earth. Presumably the Free Basics program at the heart of the problems in Myanmar and Sri Lanka falls under the Connectivity umbrella. In late August, Facebook began testing a label to show users their common interests with random people they might see in a comment thread.

The activists I have met are right: the best way to bring about change is to create public pressure for it. When Tristan and I joined forces, we could not have imagined the progress in the ensuing sixteen months. Nor did we understand that creating pressure is only the first step. Millions are aware of the problem. Far fewer understand how it affects them, why the threat to society may increase, and why they should take steps to protect themselves.

I HAVE SPENT MORE than three years trying to understand Facebook's role in the 2016 election and other undesirable events. In the process, I have learned about the other ways the internet platforms have transformed society and the economy. While intellectually stimulating, the journey has been emotionally draining. I have learned things about Facebook, Google, YouTube, and Instagram that both terrify and depress me.

This story is still emerging. As this book makes clear, I still have more hypotheses than conclusions. That said, I am convinced that Facebook's culture, design goals, and business priorities made the platform an easy target for bad actors, which Facebook aggravated with algorithms and moderation policies that amplified extreme voices. The architecture and business model that make Facebook successful also make it dangerous. Economics drive the company to align—often unconsciously—with extremists and authoritarians to the detriment of democracy around the world.

Facebook, Google, and Twitter insinuated themselves into the public square in nearly every country in which they operate and today they dominate it in many, including the United States. They have assumed a role normally reserved in democracies for government. Unlike a democratically elected government, the platforms are not accountable to their users, much less to the countries on which they have impact. To date, the platforms have not demonstrated any understanding of the responsibilities that come with control of the public square.

When Zuck first proposed his goal of connecting the entire world, that meant developed countries with broadband telecom infrastructure. The addressable population would have been less than 1.5 billion. Smartphones opened up dozens of emerging economies, increasing the potential audience to something like 4 billion. Only a few products appeal to that many people, but Facebook's core product proved to be one. Getting 2.4 billion users a month to use Facebook required brilliant execution focused on maximizing growth and driven by reducing any forms of friction. To that end, Facebook optimized for ease of connection and engagement, eliminating anything that required deliberation or might cause a user to leave the site. To maximize engagement, Facebook packaged its persuasive technologies in a design that delivered simplicity and convenience. The design of Facebook trained users to unlock their emotions, to react without critical thought. On a small scale this would not normally be a problem, but at Facebook's scale it enables emotional contagion, where emotions overwhelm reason. Emotional contagion is analogous to wildfire. It will spread until it runs out of fuel. Left unchecked, hate speech leads to violence, disinformation undermines democracy. When you connect billions of people, hate speech and disinformation are inevitable. If you operate a large public network, you have to anticipate wildfires of hate speech and disinformation.

In the real world, firefighters combat wildfires with a strategy of containment. Similarly, financial markets limit panics with circuit breakers that halt trading long enough to ensure that prices reflect a balance between facts and emotion. Facebook grew to 2.4 billion monthly users without imagining the risk of emotional contagion, much less developing a strategy for containing it. The company has failed to grasp that convenience is the reciprocal of friction, and that too much convenience, too much of "what users want," creates an environment in which emotional contagion would be an ever present danger for which the company had no answer. At some level, this is understandable. Businesses strive for efficiency and productivity; internet platforms that deliver convenience have an advantage on both counts. But businesses also have obligations to their employees, communities, and the world that the internet platforms did not meet. At the scale of Facebook, Google, and Twitter, there is no excuse for failing to prepare for emotional contagion.

The internet platforms have harvested fifty years of trust and goodwill built up by their predecessors. They have taken advantage of that trust to surveil our every action online, to monetize personal data. In the process they have fostered hate speech, conspiracy theories, and disinformation, and enabled interference in elections. They have artificially inflated their profits by shirking civic responsibility. The platforms have damaged public health, undermined democracy, violated user privacy, and, in the case of Facebook and Google, gained monopoly power, all in the name of profits.

No one working inside the internet platforms objected to these outcomes enough to take a public stand against them. Some employees at Google and Amazon have raised their voices against certain military contracts and some at Google have protested the use of arbitration in legal disputes involving employees, but none have objected to their employers' business model and business practices. To provide context, let me summarize my understanding of what I have learned on this journey so far.

First, internet platforms that we love are harming the country and the world. Those platforms we love may also be harming us without our being aware of it. In addition, there are indirect harms to all of us that result from the undermining of public health, democracy, privacy, and the economy. I do not believe the platforms are causing harm on purpose. The harm is a by-product of hyper-focused business strategies that failed to anticipate negative side effects. These are really smart people operating in a culture that sees the world through the narrow lenses of business metrics and code. They have created problems for which there may be no technology solutions.

Second, Facebook, Google, YouTube, Instagram, and Twitter have too much influence on our democracy. The 2018 midterm elections did not see a reoccurrence of the large scale, foreign interference we saw in 2016, but hate speech, disinformation, and conspiracy theories remain rampant, undermining the political discourse. We may get lucky again, but getting lucky is not a good long-term strategy. The internet platforms have consistently underestimated and misunderstood the threat from bad actors, and may do so again, despite substantial new investments in election security and efforts to combat hate speech. The country can no longer afford to take democracy for granted. I would like to think that every American will now invest the time to be informed about important issues, to vote, and to hold elected officials accountable. The country needs all the critical thinking it can muster. As citizens, we would be well served to anticipate how the algorithms of Facebook, Google, Instagram, YouTube, and Twitter try to manipulate our attention and world view.

Third, users and policy makers are far too trusting of technology. It is no longer rational to assume the best about technology entrepreneurs, companies, and products. Not because they are bad people, but rather because their incentives and culture blind them to their civic responsibilities. We now know that many technology products are unsafe. The aspects that make them unsafe are central to their economic value, which means that pressure for change must come from the outside. Consumers—who should never forget that the industry refers to them pejoratively as "users"—have enormous power, both politically and economically, should they choose to exercise it.

When a new product or technology comes to market, we should be skeptical. It is important to understand the incentives underlying new products and be selective about adoption. Before buying an Amazon Echo or Google Home, we need to read up on what it means to have a private business listening to everything we say in the presence of their devices. Even if we trust the vendor, such devices may be vulnerable to hackers. Is having a device to select our music really valuable enough to justify endless snooping into our lives? Over time, platforms will use our data in ways we cannot imagine today. Before we attach a smart TV to our home network, we should understand what the vendor will do with the data it collects. To date, the answer has been "anything the vendor wants." Among other things, privacy is the ability to make our own choices without fear.

We should be particularly skeptical about artificial intelligence. As implemented by internet platforms, AI is a technology for behavior modification with far more downside than upside. For too many companies, AI is designed to take over activities that define us: our jobs, our routine preferences, and the choice of ideas we believe in. I believe the government should insist on guardrails for AI development, licensing for AI applications, and transparency and auditing of AI-based systems. I would like to see the equivalent of an FDA for tech to ensure that large scale projects serve the public interest.

Fourth, the best way to differentiate the good from the bad is to look at economic incentives. Companies that sell you a physical product or a subscription are far less likely to abuse your trust than a company with a free product that depends on monopolizing your attention. The platforms are led by really smart, well intentioned people, but their success took them to places where their skills no longer fit the job. They have created problems they cannot solve.

Fifth, kids are far more vulnerable to screen-based technology than I ever imagined. For a generation, we have assumed that exposing kids to technology was an unalloyed positive. That was incorrect, with a high cost. I will go into this more deeply in chapter 16.

Sixth, users have no idea what is happening with their data. There is no way companies should be allowed to collect user data and then claim ownership to it. We cannot retrieve our data, but we should be able to control how it is used. Each person should know every entity that holds their data, how it got there, and how it is used.

Seventh, it is just not realistic at the scale of Facebook or Google to have the community police content. Too much harm happens during the process. Moderation can help, but has failed to date, particularly with hate speech, in part because of constraints imposed by other corporate priorities. We have asked the platforms politely to fix their problems with hate speech and disinformation. Now is the time for stronger measures.

Eighth, the culture, business model, and practices that made internet platforms spectacularly successful produce unacceptable problems at global scale that will not resolve themselves. Here again, the platforms continue to resist necessary changes. If policy makers and consumers want the problems to go away, they will have to force changes to the business model and business practices.

One business practice I want to eliminate is the use of microtargeting in political advertising. Facebook, in particular, enables advertisers to identify an emotional hot button for individual voters that can be pressed for electoral advantage, irrespective of its relevance to the election. Candidates no longer have to search for voters who share their values. Instead they can invert the model, using microtargeting to identify whatever issue motivates each voter and play to that. If a campaign knows a voter believes strongly in protecting the environment, it can craft a personalized message blaming the other candidate for not doing enough, even if that is not true. In theory, each voter could be attracted to a candidate for a different reason. In combination with the platforms' persuasive technologies, microtargeting becomes another tool for dividing us. Microtargeting transforms the public square of politics into the psychological mugging of every voter.

Ninth, I believe the threat from internet platforms justifies aggressive regulation, even with all the challenges of doing so in tech. The goal is to slow down the platforms and change incentives. The platforms have shown little ability or willingness to reform themselves, so the alternative is regulation. Congress has a lot of work to do to prepare for its role in oversight, but I believe the capability and commitment are there. Regulations are required relative to public health, democracy, privacy, and antitrust. In the near term, regulation can introduce appropriate friction to slow down the internet platforms, which is a necessary first step. Long term, it can change incentives to change behavior. I will discuss specific proposals in chapter 15.

Tenth, the country would benefit from an honest conversation about the values we expect from businesses, with a focus on the things we would sacrifice in service of those values. In the realm of technology, what will we give up to protect democracy? For example, would we sacrifice some convenience to safeguard elections? What would we forego to safeguard public health? To ensure privacy? To promote a vibrant entrepreneurial economy? A bit of friction in our relationship with technology may yield huge benefits.

Eleventh, technology has unlimited potential, but the good of society depends on entrepreneurs and investors adopting an approach that respects the rights of users, communities, and democracies. If the country and the world allow the laissez-faire capitalism that has powered the internet platforms to continue, the cost will be ongoing damage to public health, democracy, privacy, and the economy. Is that what we want? Bad outcomes are not inevitable, but we need to overcome inertia to prevent them.

Twelfth, with much reluctance I have concluded that platforms like Facebook, YouTube, Instagram, and Twitter are currently doing more harm than good. I would like to think we can clean up the mess, but we must summon the will.

WE CAN STILL HOPE that Zuck and Sheryl will eventually embrace the responsibilities that come with global dominance of an industry that influences democracy and civil liberties. Facebook has exceeded their wildest dreams, and they have made giant fortunes from it. The time has come for Facebook's leaders to accept their civic responsibilities and put users first.

The world's democracies want Facebook to act responsibly. Eventually those democracies will be able to compel change. Europe has taken the first step. It would be a smart move for Facebook and the other internet platforms to anticipate where this is going and go there without a struggle, so as to preserve value and goodwill. If Zuck and Sheryl won't take on the reform mission themselves, perhaps their employees and advertisers will rise to the challenge. Facebook's employees have shown some interest in Tristan's ideas about humane designs, which is very exciting, but we have not yet seen evidence that they have the power or will to effect the changes in business practices and business model that would be necessary to provide meaningful protection to users. Even with horrible news from Myanmar and Sri Lanka and the mounting evidence of Facebook's harm to democracy, employees have been reluctant to come forward as whistle-blowers. That is incredibly disappointing. Until change comes from within, we have to keep up the pressure on policy makers and the public. We have made a lot of progress in raising awareness, but much heavy lifting is still to come.

The most important voice can and should be that of the people who use Facebook, who will have to decide if they care more about the convenience of internet platforms or the well-being of themselves, their children, and society. It should not be a tough call. The fact that it is speaks volumes about our addiction to internet platforms.

13

The Age of Surveillance Capitalism

Who knows? Who decides? Who decides who decides?
—SHOSHANA ZUBOFF

August 2019 (One year later)

Hardly a day goes by without a new revelation about the impact of internet platforms on public health, democracy, privacy, and competition. The business model of Facebook, Google, and the advertising-supported elements of Microsoft and Amazon is metastasizing in ways that increase the threat to society. Facebook initially dominated the coverage, but attention spread to Google, Twitter, and to a lesser extent Amazon. Microsoft, whose strategy increasingly mimics a combination of Google and Facebook, has managed to avoid criticism for similar business practices. What follows are a few of the lowlights.

In October 2018, Facebook reported that hackers had penetrated its system and stolen identity tokens from twenty-nine million users, gathering extensive personal information on fourteen million of them. The hackers gained the ability to impersonate those users on other internet platforms without detection, making this by far the worst security failure at Facebook yet revealed. Facebook reported that the hackers might have been scammers. A follow-up story revealed that Facebook learned about the hack and took steps to protect its employees, without alerting users. The judge overseeing a lawsuit related to the incident expressed an openness to "bone crushing discovery" to shine a light on Facebook's business practices.

The inadequacy of Facebook's election protection measures was exposed just prior to the 2018 midterms when *Vice* attempted to place ads in the names of all one hundred US senators, the vice president, and ISIS . . . and Facebook approved them all. Facebook's WhatsApp subsidiary was blamed for hate speech and election interference in many countries and might have played an outsized role in Brazil's recent presidential election. In the 2019 European Union election, a political party with 11 percent voting support managed to gain an 85 percent share of voice on Facebook, a large multiple of all other political parties combined. If Facebook's election defenses worked as promised, the technique the party used should have failed. In addition, researchers reported that the election ad databases created by Facebook after 2016 were all but useless. As a result, a number of organizations attempted to create their own searchable databases of Facebook election ads. In one case, thousands of people volunteered to use a browser plug-in created by the public interest news organization ProPublica and to share information that Facebook provided to explain why those people were targeted. Rather than embrace research that might help protect elections, Facebook blocked ProPublica and two other firms, accusing them of privacy violations. Orwellian accusations like this one—where every party to ProPublica's research project had given informed consent—are increasingly commonplace from Facebook.

The news for Google in late 2018 was no better than for Facebook. The company missed several opportunities to cooperate with authorities in Europe, leading to a series of ever-larger penalties, the most recent of which is a proposal to change policy on copyrights in a way that might undermine YouTube. By taking the position that it is not accountable to European authorities, Google forced them to keep raising the stakes. In September, Google's CEO refused to participate in a US Senate hearing on foreign interference in elections, a hearing in which Google would have looked relatively clean in comparison to Facebook and Twitter. The empty chair at the hearing spoke volumes about Google's arrogance. The blowback appeared to trigger a new approach by Google, which soon thereafter announced it would cooperate with the second European antitrust judgment, the one that called for unbundling Google's software services from the Android operating system.

Things got really ugly for Google in October 2018, when the company announced it would shut down its failed social network, Google+. It turns out there had been a massive hack of Google+ data, which the company covered up for months. Then *The New York Times* reported that Google had paid a ninety-million-dollar severance to Andy Rubin, the co-founder of Android, despite credible evidence of sexual impropriety. This and the news that other male Google executives had escaped punishment for inappropriate sexual behavior triggered a walkout by an estimated twenty thousand employees worldwide. On the heels of smaller protests against Google bids on defense contracts, the walkout offers some hope that employees at internet platforms may eventually seize the opportunity they have to force change. Recognizing this, Google's management took steps to prevent future protests. Most recently, it announced a policy to discourage political conversations in the workplace.

Google also remains vulnerable to election manipulation, particularly through its search engine and YouTube. The sad truth is that even if internet platforms could limit overt political manipulation, they would still pose a threat to democracy. Filter and preference bubbles will continue to undermine basic democratic processes like deliberation and compromise until something comes along to break users out of them. Behavioral addiction, bullying, and other public health issues will remain.

We will also be wrestling with pervasive loss of privacy and online security. We will still be subject to manipulation. The economy will continue to suffer from the anticompetitive behavior of monopolists.

Things did not get any better in 2019. In August, the American Political Science Association gave its annual prize in political psychology to a paper titled "A 'Need for Chaos' and the Sharing of Hostile Political Rumors in Advanced Democracies," which highlighted the role that internet platforms play in enabling people who embrace "chaos incitement" to impact the political process. Rumors and disinformation are their weapons of choice. The paper argued that such people are not motivated by ideology so much as by a desire to destroy political elites. The impact of chaos inciters has been global, with no sign that it will end of its own accord.

Extremists and other bad actors continued to exploit social media—especially Facebook, but increasingly the entire constellation of internet platforms—to harm innocent people. A mass killing in Christchurch, New Zealand, which resulted in fifty-one deaths in two mosques at the hands of a white supremacist, broke new ground in the orchestrated use of social media to incite and then amplify the damage from terrorism. Prior to the attack, the terrorist organized hundreds of co-conspirators online and published a manifesto on 8chan. He livestreamed the attack on Facebook Live. His co-conspirators recorded the livestream and re-published it all over the web, including on YouTube and Instagram. It is hard to overstate the impact of this violence on the people of New Zealand. Despite huge pressure from the government of New Zealand, which made exhibiting the video illegal, internet platforms have not been able to eliminate all copies or prevent new ones from being posted. Eric Feinberg, an analyst of the deep web based in New York, assisted New Zealand in finding copies of the livestream video in the weeks and months following the attack. Since then, there have been at least two acts of mass murder in the United States perpetrated by killers who claim to have been inspired by Christchurch. In late May, an FBI field office posted a terror threat report citing the increasing threat from perpetrators activated by conspiracy theories. The algorithms of internet platforms have become a powerful amplifier of conspiracy theories, reflecting the high engagement value of such content.

In politics, *The Guardian* broke a story about allies of Boris Johnson's who used Facebook to spread propaganda in support of Johnson's campaign for prime minister of the United Kingdom. *The Guardian* revealed that the propaganda campaign exploited loopholes in Facebook's new rules for election advertising and that the agency that led the campaign has performed similar services for clients in other countries. There is every reason to believe that similar tactics may be in use in the United States.

Facebook was not the only internet platform under fire for the consequences of its business practices. In an interview at a conference in San Francisco, *The New York Times* columnist Kara Swisher criticized YouTube's CEO, Susan Wojcicki, for failing to eliminate damaging content targeted at children. The issue of inappropriate content on YouTube Kids has been around for several years, and YouTube does not appear to have made much progress. An exposé about YouTube revealed that management ignored warnings from key employees in pursuit of higher engagement and user count, leading to an explosion of hate speech, disinformation, and conspiracy theories. The former YouTube algorithms engineer Guillaume Chaslot revealed that YouTube recommended the conspiracy theorist Alex Jones more than fifteen billion times, a number that Tristan Harris noted exceeds the combined engagement of several leading news organizations. YouTube promised to do better, but a follow-up investigation by *The Verge* suggested that the situation has not improved noticeably. In September 2019, the Federal Trade Commission (FTC) and the attorney general of New York State imposed a record $170 million fine on YouTube for compromising the privacy of minors in violation of the Children's Online Privacy Protection Act (COPPA). Google's stock price rose by more than 1 percent.

After more than two years of revelations, journalists and policy makers no longer take the promises of Facebook and Google at face value. That has not prevented platform executives from framing issues to their advantage, even when doing so strained credibility to the breaking point. In early March 2019, Zuck published an essay on Facebook saying he was committed to end-to-end encryption on all of the company's platforms, including Facebook, Instagram, WhatsApp, and Messenger. Zuck framed encryption as a benefit to users, but the move would help Facebook more than users by relieving it of responsibility for inappropriate content, without necessarily improving user privacy. Later that month, Zuck published an op-ed in *The Washington Post*, arguing for new rules in four areas: reductions in harmful content, legislation to protect elections, national privacy protections modeled on Europe's General Data Protection Regulation, and regulations to ensure data portability. Where an untrained observer might see Zuck's op-ed as progress, more experienced ones saw a smart CEO trying to deflect responsibility onto governments. In my rebuttal essay in *The Guardian*, I called Zuck's op-ed "a monument to insincerity and misdirection."

Google's CEO, Sundar Pichai, might have one-upped Zuck's chutzpah in an opinion piece for *The New York Times*'s Privacy Project that positioned his company as a champion of user privacy. Statements like "Google will never sell any personal information to third parties" should not cause anyone to relax. Between its own products and data acquired from third parties, Google may have more data on each of us than any other business or even the government. No one has suggested that Google sells data. That is not the danger. The focus of critics has been on the potential for harm from exploitation of that data by Google itself. Pichai's assertion "you get to decide how your information is used" elides the deception that has come to characterize platform terms of service and user controls. It also fails to account for widespread user ignorance of the scale of data in Google's possession and the resulting lack of awareness of the need for self-defense.

Pichai's op-ed was soon followed by testimony from a Google executive that might have been textbook gaslighting. At a June 2019 hearing of a Senate Commerce subcommittee, Google's user experience director, Maggie Stanphill, responded to a question by saying, "No, we do not use persuasive technology at Google." If I had been drinking milk at the time, I would have spit it out my nose. Perhaps Ms. Stanphill was redefining "Google" to something unrelated to its legal or colloquial meanings, or perhaps she has redefined "persuasive technology"; either way, giant products like the Android operating system and YouTube contradict her, as do others.

Among the major social platforms, only Twitter has been candid about its choices. When asked why his platform did not treat white supremacy like terrorism, a Twitter employee provided a stunning justification: because doing so would require blocking some Republicans. Twitter is understandably reluctant to block politicians, even when they repeatedly violate the terms of service. There is also a business rationale. When the president of the United States generates massive value for your platform, you have an incentive to make exceptions when his tweets violate the terms of service. Doing the same for his allies almost certainly makes good business sense in the short term.

WE DO NOT THINK of Amazon as being in the same business as Google and Facebook, but that is changing. Having started as a retailer, Amazon has extended its tentacles into cloud services, advertising networks, and smart devices. Each of these is a potential gold mine for monetizable data, but with political risk similar to Google and Facebook. *Bloomberg* broke a story that Amazon employees and contractors listen to audio from Alexa and transcribe it for quality-control purposes, a reminder that once captured, data has a life of its own. We have no control over how it will be used. Subsequent stories revealed that Apple and

Google also had employees or contractors review audio from personal assistants. Both companies stopped the practice while implementing systems to get prior approval from users.

Amazon also gathers data in innovative ways. A perfect example is the competition for Amazon's second headquarters. The promise of fifty thousand jobs and a construction investment of five billion dollars attracted applications from more than two hundred cities across the United States, Mexico, and Canada, each with a detailed analysis of the local economy and plans for development, many offering tax incentives and promises of civic improvements. In the end, the two cities Amazon chose for its second headquarters were New York and Crystal City, Virginia, locations that would have been odds-on favorites from the start. By organizing a competition, Amazon was able to gather masses of proprietary data that would not otherwise have been available to it. It remains to be seen how Amazon will exploit the trove of data or if there will be any political cost for gathering it.

Amazon has also drawn criticism from civil liberties advocates for marketing surveillance technology to police departments. The company sells a facial recognition product, Rekognition, to law enforcement despite increasing evidence that such software often misidentifies innocent people. Amazon has also attempted to sell the software to Immigration and Customs Enforcement. In July 2019, Motherboard reported that Amazon's Ring subsidiary has entered into partnerships with at least two hundred law enforcement agencies related to video from its smart doorbells. In most cases, Ring offers smart doorbells to police departments in exchange for endorsements and use of Ring data in law enforcement. Ring offers homeowners an app called Neighbors that enables the sharing of video among households in a community and with police departments, all in the name of security. Ring owners have to consent to sharing with police, but once the consent is granted, no warrant would be required to obtain footage. Ring markets its smart doorbells to neighborhoods, preying on fear of strangers. Unfortunately, the program may aggravate implicit racial bias. The American Civil Liberties Union has expressed concern that in combination with facial recognition software, Ring cameras would enable widespread police surveillance without adequate protection of civil liberties. *The Guardian* subsequently reported that Ring reportedly gains power over some forms of police communication with their community and offers police departments "access to its platform in exchange for outreach to residents." Soon thereafter, Amazon CEO Jeff Bezos told reporters that he supports government regulation of facial recognition technology.

Microsoft has joined Google, Facebook, and Amazon in building a business around consumer data. Like Amazon, Microsoft has been able to avoid scrutiny of its data-related business practices, but that is not because there is nothing to see there. Products like *Minecraft* and Xbox Live capture data from children, LinkedIn and the Office 365 suite do the same for professionals, and the Bing search engine covers everyone. Microsoft's terms of service explain in easy-to-understand language that the company acquires data wherever it can and reserves the right to monetize it. Microsoft's 2018 Form 10-K to the Securities and Exchange Commission (SEC) includes a very long list of risk factors related to its gathering, retention, and commercial use of data, as well as to its applications of artificial intelligence. Tracking the changes in terms of service and risk factors over time reveals the steady incursion of Google, Facebook, Amazon, and Microsoft into what used to be the private lives of individuals, often in ways that are outside the control or consent of the affected people.

Europe has been fighting back against internet platforms, while China has generally blocked them in favor of local alternatives that further its political objectives. Against this backdrop, the US Federal Trade Commission, the Antitrust Division of the Department of Justice (DOJ), and the Securities and Exchange Commission turned their attention to internet platforms. Engagement by the FTC and the DOJ was welcome news and would have been unimaginable a year or two earlier. In July

2019, the FTC went first, with an unprecedented five-billion-dollar fine against Facebook for violating its consent decree in the context of the Cambridge Analytica privacy scandal. Investors greeted the record fine as good news, pushing the stock to within striking distance of record highs. Three factors contributed to investor enthusiasm: Facebook did not have to admit wrongdoing, it will play a significant role in managing the follow-on regulation, and the company and its management received blanket immunity for privacy violations prior to the summer of 2019. Dissenting commissioners complained that Facebook had too much influence on the regulatory process and had come out a winner. It speaks volumes about the profitability of internet platforms that investors could be happy about a five-billion-dollar fine. As a follow-on to the FTC fine, the SEC announced a hundred-million-dollar penalty to Facebook for securities law violations related to Cambridge Analytica. That same week, the FTC announced an antitrust investigation of Facebook, while the Antitrust Division of the DOJ said it would begin an antitrust review of Big Tech, specifically Google, Amazon, Facebook, and Apple. After twenty years when the only US antitrust actions related to mergers, these announcements represented a meaningful change in regulatory approach. It is not clear whether future antitrust actions will follow the lead of the industry-friendly approach of the FTC in its privacy judgment against Facebook and, if so, whether that will stimulate the passage of stronger legislation in Congress. At this time, there is bipartisan interest in such legislation but little momentum.

In September 2019, a story broke that raised the regulatory threat to Facebook and Google in a way that would have been unimaginable only a year earlier. First, a group of attorneys general from eight states and the District of Columbia announced an antitrust investigation of Facebook. The following business day, attorneys general from forty-eight states plus the District of Columbia and Puerto Rico did the same to Google. One state attorney general told me that the announcements came together very quickly, suggesting political motivations, but serious investigations may also be forthcoming. The timing remains uncertain. Facebook responded by sending Zuck to Washington, DC, to implement what CNN and others described as a "charm offensive" to the president and members of Congress. Critics in Congress remain numerous, including both Democrats and Republicans, but reports indicated that Zuck received a sympathetic ear from a number of politicians. Only eighteen months after his awkward appearances before Congress regarding Cambridge Analytica, Zuck has improved his political skills. *Gizmodo* expressed concern about the evident success of Facebook's public relations strategy of delay and deflection, which often outlasts the attention span of journalists and policy makers.

No one should be surprised at the vigorous defense of their business practices by platforms like Facebook and Google. That is their right. At the same time, though, it is our right, and perhaps our obligation, to push back. The interests of internet platforms—of Google, Facebook, Microsoft, and Amazon—do not always align with those of users or the nations in which they live. In many ways, internet platforms are the twenty-first-century equivalent of the chemical industry of the fifties. In that era, chemical companies were fast growing and enormously profitable, thanks in part to the fact that they could dump their waste products without being responsible for the cost to the environment and society. Their customers followed suit. Eventually, society decided that the industry should pay the cost of cleaning up the mess from mercury dumped in freshwater, mine tailings left on hillsides, inadvertent toxic spills, and the like. Remediation of harm reduced chemical industry margins dramatically but changed incentives for chemical companies, reclaimed the environment in many communities, and improved public health for millions. The experience with the chemical industry taught us that it is not sufficient to force cleanups; it is necessary to change incentives to prevent toxic spills in the first place.

The harm to public health, democracy, privacy, and competition from internet platforms can be viewed as toxic digital spills. Like the chemical industry before them, the platforms believe the cost of these spills should be borne by society. Unlike most toxic chemical spills, the damage from some toxic digital spills cannot be cleaned up. Is this situation fair? Why should platforms reap huge profits from the harm they cause? Should they not be responsible for the cost? If we want to limit future harm, changing incentives is essential. At a minimum, there should be a robust debate about the responsibilities of internet platforms to the people and communities they touch.

IN JANUARY 2019, a book came out that changed everything for me and many others. *The Age of Surveillance Capitalism*, by Shoshana Zuboff, does for the twenty-first century what Adam Smith's *Wealth of Nations* did for the eighteenth and nineteenth, describing for the first time the dominant economic system of its era. She analyzes the economic models of Google, Facebook, and other internet platforms, identifies the ways in which these companies have transformed the economy and society, and warns of dangers to come. It is an intellectual tour de force.

I started reading *The Age of Surveillance Capitalism* shortly after I began the tour for the hardcover edition of this book. It took me two months to finish, but the transformation in my thinking began with the first chapter. My book tour presentations incorporated Zuboff's ideas almost immediately. Just like Tristan Harris before her, Zuboff guided me through complicated issues, helping me become a better activist. At first she did so through her book, but within a few months we met and began to collaborate.

Surveillance capitalism is a business model fueled by data rather than oil. As Zuboff describes it, surveillance capitalism is a new economic system that "unilaterally claims human experience as free raw material for translation into behavioral data" that can be monetized. Where industrial capitalism uses technology to control the environment, surveillance capitalism uses technology to control human behavior. Stated another way, the goal of surveillance capitalism is behavioral manipulation for profit. Most users have no idea that manipulation is possible, much less that it is the business model of their favorite products. Individually, the manipulations may seem harmless, but collectively the cost to us as individuals and to society is enormous. We lose agency over our lives, undermining our ability to be good citizens. And this is just the beginning.

It took many years for Google, Facebook, and their imitators and enablers to make behavioral manipulation a reality, but that is where they are today. Surveillance capitalism underlies all the problems I describe in this book. It is a clear and present danger to public health, democracy, privacy, and competition. Left unchecked, it will undermine society in ways that may take generations to fix, if a fix is possible at all.

Zuboff argues that surveillance capitalism began at Google between 2000 and 2002 and evolved steadily thereafter, to the point where Google is now implementing projects designed to manipulate human behavior at city scale while also manipulating individually the choices and behavior of billions. She makes the case that Google committed itself to converting all human experience into data, which it could then use to build digital models to represent every human being. Tristan Harris describes these models as "data voodoo dolls." They can be used to predict human behavior and manipulate it. Selling behavioral predictions is a much better business than selling data or demographically targeted ads, which is why Google's commitment to retain data should not be comforting. Most of the time when you think internet platforms are listening or reading your mind, it is really the data voodoo doll at work, making behavioral predictions. Facebook followed Google's lead but did not have the full capability until 2013 or 2014. Microsoft and Amazon are making aggressive plays to catch up, while dozens of companies outside tech are hoping for a taste of the surveillance capitalism pie.

Zuboff's analysis shows how the transformation of human experience into data is the first step that enables platforms to make predictions about the future behavior of every consumer. Behavioral prediction transforms advertising by combining individual targeting with precise timing. The marketers who are Google's customers are able to buy certainty, or something very close to it, which had never previously been available at scale. Google collects and acquires oceans of data about huge populations, looking for patterns prior to major purchases or life events, such as the purchase of a car or a pregnancy. Then it tracks individual behavior, looking for the same signals. When patterns line up, the platform can predict with relatively high confidence when a person is about to buy a car or when a woman becomes pregnant. Google may be able to make a high-confidence prediction about the impending purchase or pregnancy before the buyer or mother-to-be becomes conscious of it. Google can even predict when a mother is considering a change in diaper brands. Behavioral predictions offered by Google and Facebook have transformed marketing. It is bad enough that marketers know about our life events before we do, but surveillance capitalists make that opportunity available to anyone willing to pay. Imagine the damage that would result from the promotion of the antivaxx conspiracy theory to mothers-to-be, including ones who do not yet know they are pregnant.

Perfect information about each consumer is one of the holy grails of marketing, a previously unattainable goal. Marketers will pay a huge premium to reach the right prospect at the right time. Broadcast advertising, whether it be on television, radio, newspapers, magazines, or billboards, cannot compete on either metric, so marketers have shifted ever more advertising dollars to Google, Facebook, Microsoft, and Amazon.

Consumers trust platforms to be honest brokers of information, but surveillance capitalism corrupts incentives. In reality, platforms exploit the same data voodoo dolls that give marketers perfect information to tailor search results and news feeds, potentially increasing the probability that the behavioral prediction they have sold will come true.

Advertisers get access to perfect information about consumers, but what do consumers get? They get search results and news-feed posts that serve the platforms and marketers. One core tenet of traditional capitalism is uncertainty on both sides of every transaction. That uncertainty contributes to making markets work. In surveillance capitalism, one side has near certainty, and the other has only what the partner of the people with certainty chooses for them. This information imbalance is harmful and demands investigation by policy makers charged with consumer protection. Marketers have long dreamed of manipulating consumer behavior, and surveillance capitalism enables it to a degree that is unprecedented. At the moment, no one stands in their way. None of us is immune, because the tools of surveillance capitalism apply to everyone, preying on elements of human psychology that are beyond our control.

Marketers have embraced surveillance capitalism despite serious issues. They advertise on platforms, even though huge numbers of ads are being "seen" only by bots, or on devices on racks in warehouses, fraudulently exploiting the lack of transparency in online advertising. One advertising technology expert told me that inauthentic activity may account for half the traffic to many sites. Marketers keep advertising, even though their ads may appear adjacent to posts offering illegal drugs or spreading propaganda for ISIS. The perfect information of data voodoo dolls is irresistible because it is so much easier than what came before. It is as addictive as an opioid. A major obstacle to reform is a conflict of interest relative to the advertising agencies who represent marketers. Agencies get paid a percentage of total ad billings, which gives them an incentive to protect the status quo. This may explain the reluctance of many industry participants to cooperate with an FBI investigation of media buying practices, as described in an AdAge story from March 2019.

Once they embrace surveillance capitalism, marketers are trapped. Platforms have enormous advantages over even the largest marketers.

Google, Facebook, Microsoft, and Amazon reveal less to advertisers than traditional media, but marketers have no choice but to trust them. Platforms control access to the audience that marketers need to reach. Marketers have no leverage. No individual advertiser accounts for more than a tiny fraction of revenues for any platform. Worst of all for marketers, platforms like Amazon now compete directly with the products they sell. What Google and Facebook did to journalism may now happen to many consumer product companies.

Consumers misunderstand how their relationship to internet platforms has changed. At one time, the deal was a little personal data in exchange for compelling and convenient services, paid for by targeted advertising, but that has not been the case for years. As it happens, the data consumers give up often accounts for less than 1 percent of their data voodoo doll. You read that correctly. The data you put into internet platforms represents a vanishingly small percentage of the data they have about you. The vast majority comes from other sources, including web tracking; scanning of emails, documents, and messages; third-party data from banks, credit card processors, health-care data providers, mobile carriers, affinity programs, and other apps; and data from surveillance products such as Alexa-based smart devices, Google Home, Facebook Portal, Google Street View, Pokémon Go, and Sidewalk Labs, an initiative of Google's parent, Alphabet. Everything you do these days—whether online or in the real world—leaves a data trail. Anyone who can collect it all will have an exceptionally high-resolution picture of you. A data voodoo doll. Imagine what an internet platform with that kind of information about you could do with it. For platforms like Google, Facebook, Microsoft, and Amazon, which interact with users daily, knowing everything there is to know about every person who is online enables a market value ranging from half a trillion to a trillion dollars. And that is just the beginning. The technology they use to exploit that data is likely to "improve" dramatically in coming years. And we, the people, are currently undefended. Being a digital native will not help if the manipulation of others can undermine our security or opportunities.

Zuboff's analysis shows that the incentives of surveillance capitalism require that platforms gather data everywhere it exists. They do so typically without asking permission. Consider, for example, the case of Google Street View. As Zuboff describes it, Google realized in the early years of the twenty-first century that the world was filled with unclaimed data and decided to stake a claim to as much of it as possible. With Street View, Google sent cars outfitted with 360-degree cameras up and down every street, capturing pictures of each home, including whoever and whatever happened to be visible. Google did not ask permission. It just did it. In Germany, where memories of the Stasi secret police are still fresh, a huge backlash forced Google to withdraw Street View. The rest of the world, which did not share that experience, passively accepted Street View. If Street View were the worst case, it might be appropriate to dismiss it. Unfortunately, it was just an early invasion into what used to be private spaces. Google did the same thing, but from above, with the satellite view in Google Maps and expanded from there.

But the images in Street View and Google Maps' satellite view are static. As Zuboff explains, Google wanted to convert all human experience into data, and for that it needed real-time surveillance. One early effort produced Google Glass, eyeglasses that included both a tiny computer display and a camera. Google Glass enabled users to view their computer screen all the time, but the far larger benefit accrued to Google, which captured human behavior in real time, including facial recognition of every person the user encountered. The people being recognized have no opportunity to exercise their privacy rights. Fortunately, Google Glass failed as a product, and Google withdrew it from the market.

Google engineers, scientists, and managers did not give up on the technology underlying Google Glass. Zuboff notes that they went back into the lab and repackaged the functionality of Glass as a video game, which they eventually spun out through an independent company called Niantic. The game was Pokémon Go. One billion people downloaded and played the game, which required them to wander around their community with a phone (and its camera) pointing forward. Without realizing it, users generated masses of behavioral data for Niantic and Google. As Zuboff describes it, the imperatives of surveillance capitalism drive Google to subterfuge, sleight of hand, and deception to obscure its goals and actions.

Google does not use deception because it is evil. It deceives because that is the only effective way to run the experiment. It is convinced of the virtue of its corporate mission. (In this, Google is no different from Facebook, which has repeatedly placed user safety at risk for economic gain.) If users understood what Google was planning before they fell in love with its products, they might make different choices. Zuboff argues that Google's goal with surveillance capitalism is to eliminate inefficiency and human stress . . . and to profit from doing so. By converting human experience into data, using machine learning to create behavioral prediction models, and then employing algorithms to optimize human behavior, Google has succeeded in increasing economic efficiency by removing friction from many aspects of daily life. It has generated massive wealth and arguably become the most powerful corporation on earth. But there is a side effect of Google's success. Inefficient human institutions such as personal choice and democracy may be replaced by algorithmic processes.

Pokémon Go enabled Niantic (and by extension Google) to experiment with behavioral manipulation at an unprecedented scale. If Niantic put a Pokémon on private property, would players knock on a stranger's front door? Yes, they would, even in the post-9/11 world. Would they climb a fence? Yes, they would. Would they go into a McDonald's or a Starbucks? Yes, they would. As described by Zuboff in *The Age of Surveillance Capitalism*, Niantic did a deal with McDonald's "to drive game users to its 30,000 Japanese outlets." Zuboff also noted that Starbucks decided to "'join in with the fun,' with 12,000 of its US stores becoming official 'Pokéstops' or 'gyms.'" Starbucks also created a Pokémon Go Frappuccino. Users who thought they were playing a game were also subject to what might have been the largest behavioral modification experiment ever attempted. As with so many Google products, the goals of the user and of Google are not aligned. And this is happening to 1 billion users. For comparison purposes, there are 1.2 billion people in China, of which 800 million have internet access. For all the legitimate concern about China's behavioral manipulation program, the "social credit" system, it is not yet capable of running manipulations on the scale of Pokémon Go.

Zuboff notes that the business model of Pokémon Go is based on routing traffic to the advertiser, a form of advertising known as "foot fall." In the excitement of a game, a player can be manipulated to enter a retail establishment, which does not seem like a big deal in isolation. But Google is doing the same thing—to varying degrees—in all of its products. If you use the Google suite of products, you should consider the many ways in which Google controls your choices, and perhaps your behavior. Surveillance capitalists use data voodoo dolls to optimize search results and news feeds, increasing the probability that the behavioral predictions they have sold to marketers will come true. The danger, as Zuboff makes clear, is that surveillance capitalism ultimately undermines free will, stripping people of agency over their own lives. Each concession we make to convenience lays the foundation for further incursions by internet platforms, each potentially more threatening.

Consider the examples of Google Maps and Waze. Thanks to their utility, these products are extraordinarily popular. Consumers trust them, without being aware that Google's goals sometimes conflict with their own. Consumers use map products to get from one place to another as quickly and safely as possible. Google has more complicated objectives, one of which relates to routing. Drivers who rely on Maps or Waze to guide them through long commutes are sometimes told to take a longer route. In many cases that route reflects actual traffic conditions; at other times, it may be preemptive. For Maps, Google does something called "load balancing," which is a set of techniques to keep systems running smoothly. In the context of traffic flow, load balancing anticipates bottlenecks and routes traffic around them. That means that some people may be sent on longer routes proactively in order to maximize the speed of traffic on the system as a whole. With load balancing for Maps, Google plays God. Google was not elected to or chosen for that role. It saw an opportunity and seized it. At the system level, load balancing is clearly more efficient, and in the context of Maps perhaps it is not a big deal, but it compromises the consumer's right to choose. Convenience conditions us to accept the choices platforms make for us.

Zuboff states that the imperatives of surveillance capitalism—to gather more data, to build more refined models, and to exploit those models for economic gain—require ever-greater invasions into what was once the private sphere. In an environment where there are no laws to prohibit such behavior, aggressive companies will take advantage. Common-law notions of consent—where both sides must have the same understanding of the deal—are replaced with consent based on misinformation or no information, or with processes designed to discourage the review of terms. Too often, surveillance capitalists claim data without asking for consent. Zuboff argues that the inevitable consequence of surveillance capitalism is that democracy and individual freedom lose ground to algorithmic processes.

Whether the executives realize it or not, Google's strategy implies that it believes efficiency is more important than democracy and choice. Optimizing for efficiency sounds reasonable until you realize it runs contrary to the founding principles of the United States. The thing that we lose to surveillance capitalism is liberty. If we do not bring this debate onto center stage, with the participation of every consumer and policy maker, we will sacrifice core values without being aware until it is too late. Is that what we want? My goal is to promote awareness so all consumers can choose for themselves. Perhaps they will be comfortable with Google (and other surveillance capitalists) dictating their choices. Either way, the decision should be made by consumers, not surveillance capitalists, and it should be made with as much information as possible.

Surveillance capitalists are businesspeople in pursuit of profit in a time when there are few rules to restrict their behavior. Economic incentives drive them to take initiative, assert the right to act, challenge authority to disagree, deflect criticism, adapt when necessary, but always push forward. By design, they move so rapidly that consumers and regulators have no ability to push back. By the time they react, the surveillance capitalist is well entrenched and has moved on to its next challenge to authority. The only real-time check on surveillance capitalists, the stock market, loves the business model and rewards its leading players for their aggressiveness. Investors have shrugged off every multibillion-dollar fine, confident that the rewards of surveillance capitalism will overwhelm even the largest penalty governments can impose.

First Google, then Facebook, and now others have embraced surveillance capitalism. As Zuboff shows, these companies understood that some of the required business practices would not survive close scrutiny. Google and the rest are not candid. They use product design to disguise true objectives and misdirection to divert attention away from behaviors that might be subject to criticism. Zuboff describes what she calls "the dispossession cycle." First they do what they want. Then, when they are caught doing something unacceptable, the strategy is to deny, deflect, and dissemble. In the face of persistent criticism, they eventually concede the minimum, make superficial changes, and then return to business as usual. Zuck's op-ed in *The Washington Post*, Sundar Pichai's op-ed in *The New York Times*'s Privacy Project, and the testimony of the Google user experience director that I referred to earlier are cases in point.

OVER THE COURSE OF ZUBOFF'S evidence and analysis, it becomes clear that the imperatives of surveillance capitalism pose an existential threat to democracy and the Enlightenment values upon which the United States and other democracies were built. The threat is a by-product of a successful business strategy. Few if any employees of surveillance capitalists are even aware of it. Thanks to the control privileges of their founders, internet platforms like Facebook and Google have authoritarian cultures. As is often the case with multinational corporations, surveillance capitalists have learned to align with authoritarian governments, which further undermines their statements of support for democratic values. Worst of all, their products and business practices have enabled authoritarians to disrupt and sometimes take power in countries with a history of democratic rule.

While most of the public concern about internet platforms relates to the harm to individuals, the greater threat is to society. It may be acceptable to an individual that their data be available to anyone, for any purpose—even for manipulation—but we should not be complacent about the effects on large populations. The four harms from internet platforms threaten society as well as individuals. The harm from bullying or competing for attention on Instagram goes far beyond the affected children. We are raising a generation prone to depression and even suicide. There are bound to be implications for society. The harm from filter bubbles goes beyond the fear and outrage they provoke in individuals. Filter-bubble-enabled manipulation has undermined public health around the world, enabling acts of violence and mass killings in Myanmar, the Philippines, New Zealand, the United States, and elsewhere. The correlation between online filter bubbles and domestic terrorism in the United States is well documented, including the mass shootings at the Tree of Life synagogue in Pittsburgh and a shopping mall in El Paso. The ease with which foreign actors have exploited filter bubbles and other platform tools to interfere in elections in the United Kingdom, the United States, and elsewhere should not be acceptable. It should provoke public outrage, particularly given the failure of Facebook, Google, and Twitter to respond adequately and the inability of sovereign governments to force appropriate remedies. Then there is privacy, which I define as the ability to make choices without fear. As Zuboff eloquently argues, surveillance capitalism is eliminating the sanctuaries in human life. Platforms track our every move online. They buy data from third parties. Some scan our emails, documents, and texts for economically valuable information. Smart devices capture data about our actions in the physical world. Some now listen to us in our kitchens, offices, bedrooms, and automobiles. Individuals may not care, but as a society we should care deeply. The actions of manipulated individuals affect us all.

SURVEILLANCE CAPITALISTS LIKE GOOGLE, Facebook, Amazon, and Microsoft behave as though they are exceptional. Not subject to normal rules and oversight. Entitled to make choices on behalf of billions. At the heart of Zuboff's argument is the most disturbing truth: surveillance capitalism leads to a massive disequilibrium in knowledge. A priesthood of data scientists and computer scientists have the specialized knowledge and skills demanded by surveillance capitalism, while the rest of us are left behind. In surveillance capitalism, Zuboff says, "essential questions confront us at every turn: Who knows? Who decides? Who decides who decides?" Our current situation suggests one answer for all three questions: "surveillance capitalists."

As I write this, in September 2019, Google and Facebook are

punctuating that answer with exclamation points. They are making bold moves that may transform their relationship to government and society. Google's parent company Alphabet's smart cities initiative, carried out by Sidewalk Labs, proposed a $1.3 billion plan for a twelve-acre development in Toronto's Quayside district, wired with sensors, cameras, and systems. Better data would empower those who have it to direct employees and customers in ways that would increase efficiency by reducing the friction of choice. Data enables the powerful to set rules and limit choices for economic benefit.

Superficially, the plan seemed appealing, which is why the Waterfront Toronto agency initially supported it. When Sidewalk Labs submitted its master plan in mid-2019, a clearer, different picture emerged. Sidewalk Labs, not Waterfront Toronto, would be in control. The project would not be subject to normal political processes. Accountability would be limited. Invasions of privacy and freedom of action for Sidewalk Labs would not. Public utilities and the city government would effectively be under corporate control. A small team of activists calling themselves #BlockSidewalk raised the alarm. A handful of activists from the United States joined the cause. I was one. The decision on Sidewalk Labs' Quayside plan has been delayed to enable a public debate. That is how democracy is supposed to work.

For Google's parent company, the Toronto project is the culmination of two decades of incremental advances, analogous to programs created by NASA to prepare for the moon landing in 1969. Landing on the moon required the integration of many technologies and skills, each of which had to be perfected before advancing to the next. NASA got to the moon by testing each component of the mission separately and sequentially. Propulsion into earth orbit, space walks in earth orbit, rendezvous in earth orbit, propulsion to lunar orbit, rendezvous in lunar orbit, and descent to the moon without landing all had to be executed successfully before the first landing with Apollo 11. For Sidewalk Labs, the equivalent stages included gathering static data, building data voodoo dolls, real-time data surveillance, manipulating attention at the individual level, manipulating behavior at the individual level, surveillance at the level of populations, and manipulation at the level of populations. From my perspective, the key building blocks were search, Google Maps, Street View, Google Glass, Pokémon Go, and New York's Hudson Yards. Where the trade-offs from Google Maps or even Pokémon Go might seem harmless to many, those related to Sidewalk Labs should not. The Toronto Quayside project combines the dystopian visions of two films: *Minority Report* and *The Matrix*. The twist is that Google usurps the role of government. This is what Zuboff means when she says that Google's success would displace democracy with algorithmic processes. I can imagine that fully informed people might choose such an outcome voluntarily, but to date no community has had an opportunity to make that well-informed choice. That is my immediate objective.

Authoritarian governance has had economic success stories. In Singapore and the revitalization of China over the past forty years, citizens have accepted limited control over their lives in exchange for massive gains in prosperity. In the United States, corporations have played the role of government in "company towns" for generations, often in extraction industries located far from population centers. I cannot find a precedent in the United States for a corporation replacing a democratically elected government in a diverse, heavily populated, economically prosperous environment at the scale of a city, state, or country. We should not assume such an experiment is sure to produce a happy ending.

I do not oppose "smart cities." I oppose smart cities controlled by corporations. Barcelona, Spain, has begun a smart city project with the goal the data will be owned and controlled by the residents. That is a concept worthy of consideration. The challenge is that governmental agencies lack the expertise to provide technology-powered services to their constituents, which leaves them at the mercy of corporations that do. This situation is not inevitable, but changing it would require public will and investment.

Where Google is making a play to control a community, Facebook has targeted one of the pillars of national sovereignty. Sovereign nations derive power from controlling the means of legitimate force—police and military—and the currency. Facebook has announced a project to create a new reserve currency, a product called Libra, that would challenge the monopoly of sovereign nations.

Facebook argues that Libra will enable financial transactions by people without bank accounts in emerging countries with unstable currencies. There are many countries without workable banking services for consumers. Libra would leverage the increasing penetration of smartphones to make possible digital payments in places that would never justify a bank branch. This argument is appealing. That said, it will take many years for Libra to be effective in such countries. Meanwhile, Libra may replace overnight the hyper-volatile cryptocurrencies that enable the movement of money across borders without detection or taxes, which is to say, illegally. It may increase the incentives for those with capital to bypass regulated financial systems, with potentially harmful effects for national currencies and tax revenues.

Notionally combining the best features of blockchain database technology with those of cryptocurrency, Libra pays homage to both by applying their names to different architectures. The attraction of blockchain is the decentralized database, as well as optimization for privacy and security. Initially positioned as an alternative to traditional currencies, blockchain-based cryptocurrencies like Bitcoin have generated enthusiasm, but also frustration. They have traded more like commodities—with wild swings—making them unsuitable as a store of value. Theft and lost passwords have been persistent issues for early adopters. Most market participants either trade the cryptos or employ them to avoid government scrutiny. The scale of these activities is small and to date has posed no threat to national sovereignty.

The design of Libra addresses some shortcomings of Bitcoin and other early cryptocurrencies. To reduce volatility, Libra would be backstopped by established reserve currencies, a basket of US dollars, euros, pounds, yen, Swiss francs, and others. It would centralize key aspects of the system, which would make Libra easier to use than Bitcoin and its peers. Despite these choices, Libra has features that may be problematic for those who use it. Relative to privacy, for example, it appears that at least some data may leak to Facebook and perhaps others. That said, Libra has attracted important supporters. Facebook has recruited dozens of financial institutions as partners in the project. Partners will earn a share of the float on the traditional currencies held as a backstop for Libra, which could be hugely valuable. As with smart cities, there is much to like about a cryptocurrency optimized for stability; ideally, it would be created within the existing financial system to minimize disruption.

Objections to Libra surfaced immediately. Policy makers and central banks around the world expressed alarm that a private corporation with Facebook's market power would try to create a reserve currency. Facebook responded with conciliation, saying it would work with regulators. Are banking regulators up to the task? In the context of a currency created by a private corporation, who will protect the legitimate interests of consumers and sovereign nations? This is another example of the challenges faced by the institutions of democracy in a world dominated by technology. The challenge is greater because the promoter of Libra is Facebook, a company with a terrible track record relative to public health, democracy, privacy, and competition. Given what we know, the bar for Libra—and for any other new initiative from Facebook—should be exceptionally high.

Amazon's next big play appears to be in health care, which accounts for approximately 18 percent of the US economy. The company is already a major player in medical supplies for clinics and hospitals but is quietly adding initiatives in health-care cost management, cloud services, and medically oriented services on Alexa smart devices. CNBC has speculated that Amazon is contemplating initiatives in medical records, pharmacy services, and primary care.

To date, Microsoft has not ventured far from the business categories in which it has operated for a decade or more. What it has done is imitate the business practices of Google and Facebook within those categories. Becoming a backbone for Big Data—as well as a direct player in surveillance capitalism—has made Microsoft the most valuable stock in the world. It has done so without attracting much criticism from consumers and policy makers, who generally do not appreciate the role that data has played in the revival of Microsoft's growth and profitability, much less the risk to public health, democracy, privacy, and competition from Microsoft's imitating the business practices of Google and Facebook. How long that will last remains to be seen.

It is worth taking a moment to consider the implications for the broad economy of the aggressive expansion plans of Google, Facebook, and Amazon. How will the next phase of surveillance capitalism change the economy? If corporations perceive that data has more value than physical products, factories, and equipment as a source of competitive advantage, what will happen to the value of traditional assets? In *The Age of Surveillance Capitalism*, Zuboff reminds us that the robber barons of the early twentieth century rose to prominence on the back of a similar transformation. The robber barons benefited from the commodification of work and land as labor and real estate. They built business empires in oil, transportation, and banking that provided overwhelming competitive advantage against competitors and suppliers. A domino effect began that might have disrupted the entire economy. Without the passage of groundbreaking antitrust laws championed by Teddy Roosevelt and others, the robber barons might have extracted the value from many other industries. That did not happen, because policy makers recognized the benefits of a diverse and distributed economy. In fighting antitrust legislation, the oil, transportation, and banking monopolies employed arguments that are similar to the ones we hear today from Google, Facebook, and Amazon. Those arguments proved to be unfounded then. I suspect they are unfounded now.

Could internet platforms disrupt giant industries like financial services, transportation, and health care? In the absence of government intervention, I fear that may be inevitable. Google is already going after smart devices, transportation, health care, and government. Facebook has targeted currency and financial services. Amazon's focus is on distribution, technology infrastructure, health care, and smart devices. As we watch the spread of "smart" to categories like cars and appliances, we should be mindful of the near certainty that car companies and appliance manufacturers are no better equipped to prosper in partnership with internet platforms than were the news and media industries before them. One CEO at Ford Motor Company has already lost his job following a last-minute change in strategy by Google.

Economic theory posits that competitive economies produce more value than ones dominated by monopolies, and yet almost every industry in the US economy is operating with levels of concentration not seen in a century or more. In a battle of titans, not all monopolists are created equal. Those with control of the most valuable data will have important advantages. I suspect that data voodoo dolls, in combination with high margins and pristine balance sheets, will enable Google, Facebook, Amazon, and Microsoft to disrupt just about any industry to which they devote themselves. They do not need to succeed to cause harm. When "move fast and break things" was confined to social media, it did massive damage to public health, democracy, privacy, and competition. Even if we were happy with the trade-offs in social media, why would we allow them to spread to industries on which our lives and livelihoods depend?

THERE ARE TWO IMPORTANT pieces of context for Alphabet's initiative in smart cities and Facebook's in currency. The success of neoliberalism—which posits that markets are always better at allocating resources than institutions like government—over the past fifty years had the effect of undermining the public's confidence in government institutions and other gatekeepers. Internet platforms both exploited and accelerated the erosion of public confidence with philosophies like "move fast and break things." Smaller budgets and political obstacles also lessened the ability of government institutions and other gatekeepers to deliver the services consumers expected, creating a vicious cycle. Government institutions lack the skills and budget to adapt to a technology-driven world. Into the breach leaped internet platforms, whose business models stripped the profits and authority from gatekeepers like news and media organizations, and are now preparing to encroach on services traditionally delivered by government. Consumers are left without defenders, without a clear understanding of the threat or how to protect themselves. It is no surprise that so many fear the loss of what they have and cling to any opportunity for control of their lives. I was left to wonder if this may explain a growth in seemingly unrelated behaviors in American society, including the fear of all forms of change, the fight against regulation of firearms, the efforts to protect a coal industry that is both harmful and declining economically, the enthusiasm for tattoos and body piercing, and the explosive growth of "influencers" on social platforms. People who are frustrated by having little control over their lives may do things that do not obviously serve their interests. Internet platforms magnify this problem for their own ends and by enabling others to spread hate speech, disinformation, and conspiracy theories.

Consider the impact of internet platforms on some of the important issues of our time. We know the power of internet platforms to amplify harmful content, but we think of that in the context of its impact on individuals, rather than on society as a whole. To what degree have the filter bubbles and advertising tools of internet platforms contributed to the success of climate change denial, white supremacy, gun violence, and the campaign to avoid vaccinations? The campaign to deny climate change began before internet platforms existed, but has benefited enormously from the ability to identify prospective supporters online, to indoctrinate them, and then to deploy them politically. It is hard to imagine antivaxx being more than a fringe issue in the absence of internet platforms. The same is true of lesser issues like the flat-earth conspiracy theory. White supremacy has grown dramatically in recent years, aided and abetted by the package of anonymity and communications tools offered by internet platforms. Extremists leverage the megaphone of internet platforms to recruit people and spread their message.

How much easier would it be to address issues like climate change, white supremacy, gun violence, and the campaign against vaccination if internet platforms did not exist? Eliminating the amplification of harmful content would be a blessing, but that would not be the only benefit. Internet platforms offer convenience in a social context where consumers perceive that more convenience is always better. But what of climate change? Deliveries by Amazon or DoorDash have a larger carbon footprint than less convenient alternatives. The machine learning and artificial intelligence powering internet platforms require massive amounts of electricity. They may be the largest incremental consumer of power in the country. If we are to address issues like climate change, we must be prepared to accept less convenience.

The evidence that consumers would give up democracy and freedom of choice—to say nothing of public health, privacy, and competition—for a handful of convenient web services reflects the huge gap between what consumers believe is going on and what is really happening. The internet platforms have done a masterful job of sustaining an illusion of harmlessness. They provide immensely convenient functionality without demanding monetary compensation, and package their invasions of privacy in small doses that are hardly perceptible by users. Many in the generation that has grown up since the launch of the World Wide Web—the digital natives—profess to have no concern. Having known no other environment, some digital natives dismiss fear of surveillance capitalism as the paranoia of old people. The case made by these digital natives would be completely reasonable if the impact of surveillance capitalism were limited to each person individually. Unfortunately, that is not the case. Your data may be used to harm me or others. Innocent people have died in Pittsburgh, El Paso, Christchurch, Myanmar, and the Philippines due to the actions of people negatively influenced by content on internet platforms. Harms short of death are commonplace the world over.

I share Shoshana Zuboff's view that a destructive surveillance capitalism threatens to overwhelm our economies and our democracies. At the moment, it is doing so without a fight because few people understand what is happening. This includes policy makers, most of whom trust market forces to regulate such matters. If the past few years have taught us anything, it is that Google, Facebook, Microsoft, and Amazon are unconstrained and unaccountable. They have demonstrated no willingness to self-regulate and have market power that dwarfs all would-be competitors. Restraint must come from the outside, with full participation by governments and citizens.

For countries with democratically elected governments, radical change of the sort engendered by surveillance capitalism should not happen without fully informed deliberation among and consent from the affected population. Our immediate priority should be a thorough public vetting of surveillance capitalism, its implications for democracy and personal autonomy, and the requirements for alternative visions. The first goals should be to educate the population about the reality and harms of surveillance capitalism and to solicit their support for alternative models. There are many ways to deliver the web services people value that would not undermine the foundations of society. After all, Google, Facebook, and others did precisely that before adopting the business model of surveillance capitalism. They have shown no willingness to reform their business models and have used their monopoly powers to stymie would-be competitors, smothering them in infancy, if not sooner. As a consequence, the path to protecting public health, democracy, privacy, and competition from the ravages of surveillance technology must begin with regulatory intervention, which in turn depends on a huge number of users of Google, Facebook, Amazon, and Microsoft lending their support. However, it is not enough to regulate; government institutions must learn how to deliver technology-powered services optimized for the needs of their constituents. This is why building awareness has been my focus since April 2017.

During a panel discussion in Vienna, Austria, in September 2019, the writer and researcher Evgeny Morozov confirmed my hypothesis, but went further. He characterized the newest initiatives at Google, Facebook, Amazon, and Microsoft as distinct from surveillance capitalism. Where surveillance capitalism attacks individual autonomy, projects like Sidewalk Labs and Libra go after the foundations of society. While the actors are surveillance capitalists, Morozov believes the new initiatives present a different threat than surveillance capitalism and cannot be addressed with the same remedies. The major internet platforms are exploiting deterioration in the institutions of liberal democracy throughout Western Europe and the Americas, providing valuable services in exchange for increasing control of the levers of power. The willingness of governments to work with surveillance capitalists should be taken as evidence that the institutions of democracy can no longer protect their constituents. Morozov argued that the only solution to the new threat is to rebuild the institutions of democracy, optimizing them for a world dominated by technology. Morozov has enormous credibility on this subject, as his 2011 book *The Net Delusion: The Dark Side of Internet Freedom* sounded the alarm five years before I understood it. I found his argument to be persuasive. If we are to protect society, the best approach may be for government to attack the foundation of surveillance capitalism, while freezing new initiatives by internet platforms. These actions would buy time to rebuild the institutions of liberal democracy, which might take a generation.

THE AGE OF SURVEILLANCE CAPITALISM hit me like an avalanche. A few of the ideas had already occurred to me, but the comprehensiveness of Zuboff's analysis—and the evidence she offered to support it—transformed my understanding and eliminated doubt. Where I had been looking at Tristan's addiction hypothesis as the root cause of the damage by internet platforms to public health, democracy, privacy, and competition, Zuboff's book revealed that addiction is a tool of surveillance capitalism. The best way to address it—and to restore public health, democracy, privacy, and competition—would be to attack surveillance capitalism itself, to eliminate the means and consequences of behavioral modification. This means attacking the data economy. We have to use every tool in the regulatory tool kit and then invent some new ones. That won't fix the problems caused by addiction to internet platforms or the inability of government institutions to deliver solutions in a technology-dominated world, but it will create a better environment for doing so.

Not every technology company has jumped on the surveillance capitalism bandwagon. Apple has access to extraordinarily valuable data, but its products increasingly reflect a commitment to protecting privacy. Apple Maps does not store routes and takes steps to ensure Apple cannot reconstruct your routes. Apple's Siri does a significant amount of processing on the phone to minimize data leakage into the cloud. The 3-D images used for Apple's Face ID never leave the phone. Apple Pay makes some purchases equivalent to buying with cash. Apple effectively killed Facebook's Onavo virtual private network, which had been used to spy on user behavior, by banning it from the App Store. Apple then suspended both Facebook and Google for a day when they violated App Store rules in offering "research" products designed to track every action taken by users. The Apple Card credit card has been designed to minimize the amount of data that can be captured by third parties. In combination with Apple Pay, the card promises to break new ground on privacy in digital commerce. Apple also announced Sign In with Apple, which is designed to minimize the ability of websites to track Apple customers.

A few startups have followed Apple's lead, including DuckDuckGo and Brave with privacy-oriented browsers, DuckDuckGo with private search, and Ghostery and Disconnect with tools that blocks apps from tracking user behavior. Slack, which competes with Microsoft in online office automation, offers terms of service that provide some hope of privacy protection. That said, there are situations where data may not be protected, the most glaring of which is if the company were acquired, which is likely at some point. Given Microsoft's surveillance capitalism proclivities, Slack would be well advised to build and market its privacy advantage.

I am optimistic about our chances. On my book tour, I discovered that my issues are not partisan. I can share the same message on Fox News, MSNBC, Fox Business, CNBC, conservative talk radio, and NPR and get positive feedback in every case. I can work with the Federal Trade Commission and the Antitrust Division of the Justice Department, both elements of the Trump administration, and with members of Congress on both sides of the aisle. I have also met with members of four different political parties in Canada and two in the United Kingdom. These people may not always have identical priorities, but the underlying issues are the same for all. In a country as polarized as the United States, it is remarkable to find an issue that unifies so many policy makers and voters. On one side of the issue are the roughly one million people who work for Google, Facebook, Amazon, and Microsoft. On the other is everyone else, nearly 330 million Americans. That alone should give us hope of a favorable outcome.

The remaining chapters will address solutions. Chapter 14 summarizes where we are and our options for going forward. Chapter 15 describes what government can do. Chapter 16 does the same for you. The epilogue will wrap things up.

14

What Is to Be Done

The problem with Facebook is Facebook.
—SIVA VAIDHYANATHAN

Google and Facebook started up without modesty or a sense of irony. Google's original 1998 mission statement was to "organize the world's information and make it universally accessible and useful." Not to be outdone, Facebook launched with this mission: "To give people the power to share and make the world more open and connected." It is demonstrably true that both companies succeeded on their own terms, so much so that Microsoft, Amazon, and others are scrambling to catch up. Unfortunately, the mission statements defined success narrowly, in terms that created massive wealth for their founders while imposing negative side effects on far too many countries and consumers.

By pursuing strategies of global domination, Google and Facebook exported America's twin vices of self-centered consumerism and civic disengagement to a world ill-equipped to handle them. Tools that allow users to get answers and share ideas are wonderful in the ideal, but as implemented by Google and Facebook, with massive automation and artificial intelligence, they proved too easy to manipulate. Users believe Google and Facebook represent reality, but they are more like fun-house mirrors. Google's ability to deliver results in milliseconds provides an illusion of authority that users have misinterpreted. They confuse speed and comprehensiveness with accuracy, not realizing that Google results may be influenced by data voodoo dolls and advertiser needs, as well as search engine optimization and manipulation. Users mistakenly believe their ability to get an answer to any question means they themselves are now experts, no longer dependent on people who actually know what they are talking about. That might work if Google did not imitate politicians by giving users the answers they want, as opposed to the ones they need. Google's You-Tube does something similar, but more extreme, in video. Though it might have started as industry snark, "three degrees of Alex Jones" is a directionally accurate reflection of the way YouTube recommends conspiracy theories. Facebook has leveraged our trust of family and friends to build one of the most valuable businesses in the world, but in the process it has aggravated the flaws in our democracy—and those of our allies—while leaving citizens ever less capable of thinking for themselves, knowing whom to trust, or acting in their own interest. Bad actors have had a field day exploiting Google and Facebook, leveraging user trust to spread hate speech, disinformation, and conspiracy theories, to suppress voting, and to polarize citizens in many countries. They will continue to do so until we, in our role as citizens, reclaim our right to self-determination.

You would think that users would be outraged by the way that internet platforms like Facebook, Instagram, Google, YouTube, Microsoft, and Twitter have been used to undermine democracy, human rights, privacy, public health, and innovation. Some are, but most users love what they get from internet platforms. They do not want to believe that the same platforms that have become a powerful habit are also responsible for so much harm. That is why I joined with Tristan, Renée, and Sandy in 2017 to connect the dots for users and policy makers.

Facebook remains a threat to democracy. The same is true of the other platforms. Democracy depends on shared facts and values. It depends on communication and deliberation. It depends on having a free press and other countervailing forces to hold the powerful accountable. Facebook, Google, and Twitter have undercut the free press from two directions: they have eroded the economics of journalism and then overwhelmed it with disinformation. On internet platforms, information and disinformation look the same; the only difference is that disinformation generates more revenue, so it gets much better treatment. To platforms, facts are not an absolute; they are a choice to be left initially to users and their friends but then magnified by algorithms to promote engagement. In the same vein, algorithms promote extreme messages over neutral ones, disinformation over information, conspiracy theories over facts. Every user has a unique experience and potentially a unique set of "facts." Like-minded people can share their views, but they can also block out any fact or perspective with which they disagree.

When it comes to democracy, internet platforms do a few things well. They enable communication of ideas, as well as the organization of events. We have seen Black Lives Matter, the Women's March, Indivisible, and the March for Our Lives all leverage Facebook to bring people together. The same thing happened in Tunisia and Egypt at the start of the Arab Spring. Unfortunately, the things platforms do well are only a small part of the democracy equation, and they are easily coopted by the powerful to harm everyone else.

According to Stanford professor Larry Diamond, there are four pillars of democracy:

1. Free and fair elections;
2. Active participation of the people, as citizens, in civic life;
3. Protection of the human rights of all citizens;
4. Rule of law, in which the laws and procedures apply equally to all citizens.

When it comes to free and fair elections, we know the grim story. The platforms have made changes, but the elements that allowed interference remain and could be exploited by anyone.

Relative to active participation of the people as citizens, three key factors are the willingness to respect other viewpoints, engage with them, and compromise. When the three factors are present, deliberation enables democracies to manage disagreements, find common ground, and move forward together. Unfortunately, everything about the architecture and design of internet platforms undermines deliberation.

The third pillar—equal rights for all—should be beyond the scope of internet platforms like Facebook, but they exert significant influence. Facebook has become the closest thing many countries have to a public square. The algorithms of platforms like Facebook have more influence over our daily life than the law. Every interaction is governed by the terms of service of a private corporation whose priority is profits. Facebook applies community standards in each country—sometimes in English, rather than the language of the affected people—but in most countries those standards reflect the interests of the powerful. Enforcement of community standards is automated, with rules that can be gamed, which works disproportionately to the benefit of those in power.

The fourth and final pillar of democracy is the rule of law. Here, the impact of internet platforms is indirect. Each platform has its own "laws," which it enforces selectively. This can have an impact in the real world. For example, the use of Facebook to discriminate in housing, in violation of the Fair Housing Act, persisted long after Facebook claimed to have stopped people from doing it, as demonstrated by ProPublica. The outcome of the US presidential election of 2016, the one in which the Russians used Facebook to interfere, has resulted in a broad assault on the rule of law. So long as the business model of surveillance capitalism persists, the threat to democracy will remain.

Since I completed the hardcover edition of this book, Facebook and Google have expanded their ambitions in ways that present new threats to democracy. Democratic institutions do not have the expertise and resources to protect the interests of constituents in a world dominated by technology. Many governments are willing to turn over essential functions to corporations whose incentives are not aligned with constituents, with no clear understanding of the true cost.

Facebook remains a threat to the powerless around the world. The company's Free Basics service has brought sixty emerging countries into the internet age but at the cost of massive social disruption. Cultural insensitivity and lack of language skills have blinded Facebook to the ways in which its platform can be used to harm defenseless minorities. This has already played out with deadly outcomes in Sri Lanka and Myanmar. Lack of empathy has caused Facebook to remain complacent while authoritarians exploit the platform to control their populations, as has occurred in the Philippines and Cambodia.

Facebook remains a threat to public health. Users get addicted. They get jealous when friends show off their beautiful lives. They get stuck in filter bubbles and, in some cases, preference bubbles. Facebook helps them get into these states, but it cannot get them out. Preference bubbles redefine identity. To break through them will almost certainly require human intervention, not technology. And Facebook is not alone when it comes to harming public health. YouTube, Instagram, Snapchat, texting, some video games, and a range of other applications can undermine public health every day.

Facebook remains a threat to privacy. Along with Google, Microsoft, and Amazon, Facebook has surveillance systems that would make an intelligence agency proud. The same cannot be said of Facebook's handling of data. Both Google and Facebook subvert the intent of privacy efforts like Europe's GDPR with option-dialogue boxes designed to prevent users from taking advantage of their new rights. Smart devices powered by Alexa and Google Assistant threaten to invade whatever sanctuaries remain in our lives.

Facebook remains a threat to innovation. Facebook, Google, Microsoft, and Amazon enjoy the privileges of a monopoly, with network effects on top of network effects, protective moats outside protective moats, with scale advantages that make life miserable for would-be competitors and innovators.

The time has come to accept that the flaws of internet platforms outweigh their considerable benefits. Zuck and the team at Facebook know that people are criticizing them, and while they do not like it, they are convinced the critics do not understand. They believe connecting 2.4 billion people in a single network is so obviously a good thing that we should allow them to get back to work without further discussion. They cannot see that connecting so many people on a single network drives tribalism and, in the absence of effective circuit breakers and containment strategies, has provided dangerous power to bad actors. The corners Facebook has cut have enabled great harm and will continue to do so. If anything, Instagram and YouTube are more dangerous than Facebook. Both target young people, and neither has demonstrated an ability to protect users from harm.

The time has come to shift the focus of technology from exploiting the weakest links in human psychology to a commitment to serving the most important needs of users. Working together, Silicon Valley and users have the power to make technology a bicycle for the mind once again. Technology can provide answers and entertainment while respecting the mental and physical well-being of consumers, as well as their privacy and rights as citizens. Doing so would be a huge opportunity. Let's take a page from renewable energy, where we are addressing a man-made catastrophe by creating new industries around solar and wind, among other things.

What would this Next Big Thing look like? It does not mean losing the functionality we like today. There is a way to deliver every internet service that consumers like in a form that empowers rather than harms. The players would require a different business model, perhaps based on subscriptions. Given the ubiquity of internet platforms, it might be appropriate to treat some applications as public utilities operated for the benefit of the community. Government subsidies may be appropriate in some cases. The government already subsidizes energy exploration, agriculture, and other economic activities that the country considers priorities, and it is not crazy to imagine that civically responsible internet platforms may be very important to the future of the country. The subsidies might come in the form of research funding, capital for startups, tax breaks, and the like. For example, governments might offer large prizes to encourage the development of internet platforms with business models that do not invade privacy or manipulate attention and behavior. Imagine what would happen if the prize were $1 billion for the first company to reach a milestone like one hundred million weekly users for an alternative to Facebook, Instagram, or YouTube. Such an incentive would leverage market forces in a way that traditional government contracts might not.

Incentives to empower consumers would lead to new applications of existing technologies. Consider mouse tracking. Google and others track mouse movement to determine if a user is human and potentially to capture specific characteristics. The incentives of surveillance capitalism are such that if mouse tracking were to enable today's platforms to identify the first symptoms of a neurological disorder like Parkinson's disease, they might sell that diagnosis—that certainty—to the highest bidder, perhaps an insurance company, which might raise rates or terminate coverage. In a world that rewarded the empowerment of users, mouse tracking would enable an insurance business that alerted users to symptoms and advised them where to get medical help.

The Next Big Thing offers opportunities to rethink the architecture of the internet, pushing more processing and data storage out of a centralized cloud and onto devices and smaller clouds at the edge of the network. Distributed architectures can be relatively more secure and private. For example, I would like to address privacy with a new model of authentication for website access: a private mode. Every site and every browser would support private authentication, generating strong passwords for every log-in. The app would store private data on the device, not in the cloud. The system would pass along only the minimum information required to prove identity, which might not even include names or other personal data. For example, the service could enable logins to a media site that confirmed that the user was a subscriber, without revealing identity. The app could offer a range of log-in services from "anonymous," as might be appropriate for a content site, up to something with significant data attached, which would enable financial transactions. The goal is to make privacy the default, giving users total control over any data transfer and ensuring that data would go only to the people who actually need to have it. Platforms and merchants will be unhappy to lose access to data from users who choose private log-in, but that is their own fault. They should not have abused the trust of users. At this writing, Apple has announced Sign In with Apple, a service analogous to Facebook Connect and Sign In with Google, but sharing many of the features I just described. Every product in the market today could empower consumers if vendors chose to do so, as Apple is doing. Treating consumers as customers, rather than fuel, would enable products that do not exist yet. The market opportunities are huge. The technology is within reach. All we need is the will to pursue it.

The Next Big Thing should also include smartphones that are less addictive and do not share private data, devices in the Internet of Things that are respectful of data privacy, and applications that are useful and/ or fun without causing harm. While many smartphones have created apps to help consumers manage their usage, we are still stuck with designs that promote behavioral addiction. The vendors of smart speakers, appliances, and automobiles appear to be committed to grabbing a sliver of surveillance capitalism, without regard to consumer privacy. There are currently no rules to protect consumer privacy when next-generation wireless (5G) comes to market to power the Internet of Things (IoT). The magic of 5G is not going to be more bandwidth to phones; it will be in enabling pervasive 4G-level bandwidth at one-tenth the cost. There are other issues with 5G, specifically the national security implications of telecommunications infrastructure manufactured in China at a time when tensions are rising with that country.

For good or ill, IoT already exists. The design of first generation of smart devices does not prioritize consumer welfare relative to privacy and security. Critics have expressed alarm about the ability of Amazon's Alexa and Google Home to snoop on users. With 5G, sensors and smart devices will enable internet platforms to gather data from a much greater array of devices, most in places once viewed as sanctuaries. There is no reason to believe IoT will be benign unless users and policy makers demand it. The first generation of devices—flat-screen TVs, for example—has been plagued by data privacy issues. No one knows what is happening with the data they collect. There is no excuse for not taking steps in advance to limit harm when IoT and platforms converge.

Unfortunately, technology can take us only part of the way to a solution. We are a polarized country with really poor civic engagement. At least a third of the US population identifies with ideas that are demonstrably untrue, and a far larger number have no regular interaction with people who disagree with them or have a radically different life experience. I do not see how technology can fix that. Somehow we have to change our culture to make civic engagement a priority. If Black Lives Matter, the Women's March, Indivisible, the March for Our Lives, the increase in labor actions by teachers, and the million-person protests in San Juan are any indication, the process has already begun. Unfortunately, these important efforts in activism address only a portion of the political spectrum, and their success to date has hardened resistance on the other side.

This book has focused on Facebook because that company played the biggest role in the Russian interference in the 2016 election and by chance I tripped over that story early on. In the context of US elections, I still worry more about Facebook and its subsidiary Instagram than other platforms, but not by much. Google, YouTube, and Twitter can all be manipulated. The interference playbook from 2016 is out there. Anyone anywhere can run it in any election at any level. The Cambridge Analytica data set and possibly others like it are out there. Anyone can buy access to them on the dark web. But you don't have to go to the dark web to get detailed data about American voters. Huge amounts of data are available. Campaigns can buy a list of two hundred million voting-age Americans with fifteen hundred data points per person from a legitimate data broker for seventy-five thousand dollars. Commercial users have to pay more, but not that much more. Think about that. Commercial data brokers do not sell lists that have been paired with voter files, so it would take some effort to replicate the data set created by Cambridge Analytica, but it can definitely be done by any sufficiently motivated party. A data set that includes Facebook user IDs gets access to the latest user data every time it is used inside Facebook. Anyone can still create inauthentic Facebook Groups and accounts on other platforms to deceive voters.

With 2020 looming, we should expect efforts to influence elections through hate speech, disinformation, and conspiracy theories on every platform, including Facebook, Instagram, Messenger, WhatsApp, Google, YouTube, and Twitter. The good news is that American voters learned from 2016. While reports indicate an increase in disinformation in 2018, the three groups most effectively suppressed in 2016—suburban white women, people of color, and idealistic young people—turned out in large numbers in the midterms. Unfortunately, there are many other ways to mess with an election. It does not cost much in comparison to the potential value. Clint Watts, the national security consultant for the FBI, framed it this way in an email to me: "Many will project the past onto the future, expecting Russian disinformation to again seek to influence US elections. The Kremlin, rather than being the dominant social media manipulator globally, will be one of many seeking to surreptitiously move audiences to their preferred position using social media influence. Future political manipulators will copy the Kremlin's information warfare art but apply an advanced level of technology, artificial intelligence, to sway audiences through rapid social media assaults. The danger to democracy will not be just authoritarians, but all political campaigns and public relations firms employing social media influenced to drive audiences apart online and drive constituencies apart at the ballot box."

Zeynep Tufekci, the UNC scholar who is one of the world's foremost experts on the impact of emerging technology in politics, has observed that internet platforms enable the powerful to effect a new kind of censorship. Instead of denying access to communications and information, bad actors can now use internet platforms to confuse a population, drowning them in nonsense. In her book *Twitter and Tear Gas*, Tufekci asserts that bad actors are "inundating audiences with information, producing distractions to dilute their attention and focus, delegitimizing media that provide accurate information (whether credible mass media or online media), deliberately sowing confusion, fear, and doubt by aggressively questioning credibility (with or without evidence, since what matters is creating doubt, not proving a point), creating or claiming hoaxes, or generating harassment campaigns designed to make it harder for credible conduits of information to operate, especially on social media which tends to be harder for a government to control like mass media." Use of internet platforms in this manner undermines democracy in a way that cannot be fixed by moderators or AI searching for fake news or hate speech.

My journey of discovery began with a focus on elections but eventually included public health issues, data privacy, and the suppression of competition and innovation. My goal has evolved from starting a conversation to building awareness of problems that are complex and nuanced so that people can make well-informed choices about the role of technology and the future of the country. Internet platforms pursued unlimited scale without understanding, much less preparing for, unintended consequences. Their efforts generated unfathomable wealth at a huge cost to society, which is left with a mess to clean up. We can no longer expect the platforms to reform themselves.

This is not a conclusion to which I came easily. Silicon Valley is my world. Technology has been my life's work. Professionally, I spent thirty-four years being a technology optimist. Then came the 2016 election. Suddenly I saw things that were incompatible with my historically rosy view of technology. The more I learned, the worse it looked, until I finally realized that internet platforms had forgotten the prime directive of technology: to serve the needs of humans. Having been launched at the moment when the constraints on technology disappeared, the platforms confused easy success with merit, good intentions with virtue, rapid advances with value, and wealth with wisdom. They never considered the possibility of failure and made no preparations for it. The consequences have been profound. For the Rohingya in Myanmar, the Muslims in Sri Lanka, and some victims of gun violence in the United States and New Zealand, they can be deadly.

It is hard to accept that great harm can come from products we love—and on which we have come to depend—but that is where we are. Our parents and grandparents had a similar day of reckoning with tobacco. Now it's our turn, this time with technology. The issues go beyond internet platforms to include smartphones, texting, video games, artificial intelligence, smart devices, and other products that replace human interaction with virtual alternatives. As much as we would like an easy fix, there is none that allows us to coexist with these products in their current form.

We put tech on a pedestal. That was a mistake. We let the industry make and enforce its own rules. That was also a mistake. We trusted internet platforms not to hurt us or democracy. That was a disastrous mistake that we have not yet corrected. Without a change in incentives, we should expect the platforms to introduce new technologies that enhance their already-pervasive surveillance, behavioral prediction, and behavioral modification capabilities. Those new technologies will offer convenience and perhaps other benefits to users, but recent experience suggests we should not assume that they will be benign. As technology advances, we should expect risk and potentially harm to increase as well. The challenge is made more difficult by the rapid evolution of technology and the culture that produces it. The current wave of artificial intelligence, virtual reality, deep fakes, 5G, and smart devices are being created in the same cultural environment as internet platforms.

Even if platforms abandon surveillance capitalism and reform other business practices, they are in no position to repair the damage they have caused. With a shocking percentage of the country's population stuck in preference bubbles that blind them to fact, we need to think about ways to reconnect people in the real world, to encourage handshakes and eye contact with people who live differently and hold different views. But we should not let the platforms off the hook. We should not trust them until they earn it. We should not tolerate election interference. Business models should not be untouchable. No corporation should be entitled to undermine society without severe consequences. Governments should be prepared to shut down internet platforms to force reform.

One hopes that before long the people who run the major internet platforms will have an epiphany, but that should have happened a year or more ago. Given the damage to democracies in Europe and North America that resulted from the exploitation of their platforms, Facebook, Twitter, and Google might also consider how long they want to undermine the country in which they live.

The US economy has historically depended on startups far more than other economies, especially in technology. If my hypothesis is correct, the country has begun an experiment in depending on monopolists for innovation, economic growth, and job creation. If I consider Google, Amazon, and Facebook purely in investment terms, I cannot help but be impressed by the brilliant way they have executed their business plans. But the health of the economy is at stake. These companies do not need to choke off startup activities to be successful, but they cannot help themselves. That is what monopolists do. Why would the country take that risk?

Faced with conclusive evidence that internet platforms have a dark side, we must do something about it. As a capitalist, I would ordinarily favor letting markets settle such problems, but the market is not getting the job done. The internet platforms have disrupted many industries, including music, photos, video, and news. They have disrupted the world of tech startups, too. They have created massive wealth for their investors but little in the way of traditional economic value, normally measured in the form of jobs and infrastructure. I don't want to penalize success, but if that is the only way to protect democracy, privacy, public health, and competition, so be it. Allowing the internet platforms to disrupt every sector in the economy may not be in the national interest. People need jobs, and internet platforms don't create enough of those. Worse still, too many of the current generation of technology companies derive their value from reducing employment elsewhere in the economy. I would like to think that Silicon Valley can succeed without killing millions of jobs in other industries. In the mid-seventies and the eighties, when the United States first restructured its economy around information technology, tech enabled companies to eliminate layers of middle management, but the affected people found jobs in more attractive sectors of the economy. That is no longer the case. The economy is creating part-time jobs with no benefits and no security— driving for Uber or Lyft, for example—but not creating jobs that support a middle-class lifestyle, in part because that has not been a priority. Teaching everyone to code is not the answer, because coding will likely be an early target for automation through artificial intelligence.

I see no easy solution to the problems posed by Facebook and the other internet platforms. They are deeply entrenched. Users trust them, despite an abusive relationship. To contain the problem somewhere near current levels, we need government intervention in antitrust, consumer protection, and trade policy. I will address the opportunities for government action in the next chapter.

15

What Government Can Do

*This technology that you have invented has been amazing. But now it's a
crime scene. And you have the evidence. And it is not enough to say that
you will do better in the future. Because to have any hope of stopping
this from happening again, we have to know the truth.*
—CAROLE CADWALLADR

*My question to everybody else is, is this what we want: to let them
get away with it, and to sit back and play with our phones,
as this darkness falls?* —CAROLE CADWALLADR

The story of internet platforms is like a horror movie. It began well,
full of promise, before transforming into a series of nightmares. At
this point, government must intervene. Although there are issues with
the architecture of the internet that require attention, the core problem
is not technology. It is the culture, business model, and business practices of companies like Facebook, Google, Amazon, and Microsoft. The
success of these companies is not simply a function of genius and great
execution. It is the product of both market and regulatory failures that
have enabled four companies to accumulate unprecedented influence on

our economy, on our democracy, and on our daily lives. Influence that once seemed benign has morphed into something dangerous. And yet the country does not insist on change, despite repeated and increasingly serious infractions. If someone had asked the American people of my parents' generation—the one that fought World War II—what the penalty would be for a company that enabled a hostile power to interfere in a presidential election, I imagine that a large percentage would have advocated for a corporate death penalty. Many would have insisted on criminal prosecution of executives. The same might have been true for a company that enabled a "classic ethnic cleansing" or the recruitment and radicalization of terrorists. Why do we allow such business practices to go unpunished now? Why would we let them take control of services traditionally supplied by the government in exchange for a "smart cities" label?

Our culture has changed dramatically over the past fifty years, nowhere more so than in business. Where the management guru of the sixties Peter F. Drucker taught executives to balance the interests of five stakeholders—shareholders, employees, the communities where employees lived, customers, and suppliers—businesses today focus almost exclusively on shareholder value. That focus makes it acceptable for corporations to lay off thousands, ship jobs overseas, and engage in business practices that would have been off-limits fifty years ago. As a result, the past decade has been a golden age for investors. For the rest of the population, the costs are mounting, while the benefits are elusive.

Where government set and enforced the rules of business in Drucker's era, today's businesses enjoy a freedom from heavy oversight unseen in a century. We allowed the banking industry to undermine the global economy in 2008 without material punishment. As one journalist who covered the story told me, the Obama administration had an opportunity to shut down a major bank to set an example but chose not to do so. What level of malfeasance would be required to justify a corporate death penalty? It is a question worth asking, because in the current environment corporations are taking advantage of the lack of accountability. Once-unthinkable business practices have become normalized. This book is about Facebook and the other internet platforms, but many of the issues are pervasive in the economy. Fixing the internet platforms may be the first step to restoring balance to our economy and democracy.

Only the government can play the role of setting and enforcing the rules of capitalism. Only the government can protect consumers from harm. The time has come for government to do that.

As citizens, we must do what is necessary to rebuild the institutions of democracy so that our interests will be protected. To implement change, governments require the support of the governed. Too many consumers remain indifferent to the dark side of social media, reflecting the complexity of the problem and the effectiveness of industry communications. The good news is that consumers are increasingly aware of and concerned about the behavior of internet platforms. Politicians are listening, but activists like me still have work to do to help consumers understand both the catastrophe wrought by internet platforms and the opportunities available to us to fix the mess. Another important constituency for politicians, investors, remains enthusiastic about the culture of business in this country. They view record-breaking fines against Facebook and Google as an acceptable cost of business. They appear to be untroubled by the well-documented harms of internet platforms and have demonstrated enthusiasm for next-generation initiatives like Alexa, Sidewalk Labs, and Libra, despite the risks to society and sovereignty. Wall Street is going to hate any true reform of internet platforms or of the culture of American business. So be it. The time has come to prioritize the rest of the population. Wall Street will be fine. If we can force reform and trigger a new generation of technology companies whose products empower users, the Next Big Thing will be bigger than the previous big thing. It happens with every new tech cycle. Investors will embrace it. Greed will win out.

We need government intervention because internet platforms have far less to fear from market-based alternatives than earlier generations of tech giants. So long as they do not attack each other, the Big Four should be able to maintain their dominance for many years to come. In that context, the threats to public health, democracy, privacy, and competition will persist until consumers and policy makers commit themselves to forcing change. How best to do that? That is the goal of this chapter.

I am convinced that there is a path to happiness that eliminates what is wrong with internet platforms while preserving the good. To follow that path, we must frame the problem, evaluate our options, and commit ourselves to action. For me, a key insight since publication of the hardcover edition is that we should treat the harm to public health, democracy, privacy, and competition as symptoms of a common disease, rather than separate ailments. I have shifted my attention upstream, where the root causes are. The problem starts with the very human tendency to prioritize convenience over almost everything else. Internet platforms have deceived us about the cost of the convenience delivered by their apps. They position convenience as a feature, but it is really a bug. The pursuit of convenience makes us vulnerable to exploitation. The sooner we recognize that, the easier it will be to fix the problems enabled by it. In chapter 16, I will address the ways each of us can be part of the solution.

The challenge for government will be to identify solutions that address the interrelated elements of the problem: our desire for convenience, our vulnerability to persuasive technology, the culture of Silicon Valley, the business model of surveillance capitalism, the lack of alternatives to surveillance capitalism in the world of internet platforms, and the challenge to civic institutions from pervasive technology. So far as I can see, there is no single fix for all six issues. Fortunately, we do not need six separate solutions either. Surveillance capitalism is the key; it appears to be inherently harmful to far more than consumer privacy. For example, the incentives of surveillance capitalism compel ever more persuasive technology, so forcing changes to it may be beneficial for public health and democracy. It may help us reduce the volume of hate speech, disinformation, and conspiracy theories infecting our public discourse. The toxic culture of Silicon Valley today may also be a by-product of surveillance capitalism. The experience of the past two years suggests that these problems will not fix themselves. If we can reform surveillance capitalism, we may be able to buy time to address the other issues.

While this book has primarily focused on the harms to individuals and society, Facebook, Google, Amazon, and Microsoft also take advantage of other businesses, and no one has the power to fight back. Media companies have invested huge sums in creating content, but internet platforms frequently monetize it without a comparable investment or appropriate compensation to the creators. This has undermined the economics of news and music while threatening to do the same in portions of the video industry. Marketers, whose ad dollars generate most of the profits for internet platforms, complain constantly—and with evidence—about overstated metrics of reach and views, but have been unable to force change. Marketers have yet to recognize the threat to their business from vertical integration by internet platforms. Amazon has taken the early lead on this, implementing a strategy to reduce brand choices to two: price and convenience. It exploits its data advantage to clone the bestselling products for its own AmazonBasics line, to which it may give preferential treatment in search results. Amazon also has not prevented an explosion of brand counterfeits in its marketplace. The effect of AmazonBasics and counterfeits is to undercut the brand attributes of a wide range of consumer products. In addition, Amazon's increasing success in product search conveys extraordinary power to undermine brand advertising by favoring products consistent with its own priorities. Brand owners and advertising agencies have no way to compete with Amazon's control of the world's largest online marketplace and access to data voodoo dolls. Facebook appears to be positioning itself to join Amazon in transforming the market for consumer products. If Facebook succeeds in bringing Libra to market, it will have a payment system that in combination with its own marketplace would enable a different assault on traditional brands.

Having reluctantly concluded that government intervention is our only option, I have gone to Washington to recommend a three-pronged approach: enabling competition and new business models, attacking the foundations of surveillance capitalism, and penalizing business practices that undermine society and democracy. There may also be an opportunity to apply securities law to restrain the growth of surveillance capitalism, at least temporarily. I will address these opportunities in turn.

AMONG POLICY MAKERS, there is growing support for antitrust intervention. The machinery of antitrust regulation in the United States has been idle for a generation. Even at full speed, antitrust investigations and prosecutions always took many years, sometimes a decade or more, to complete. The good news is that even an investigation by the DOJ or the FTC can sometimes change behavior for the better.

The headlines about US antitrust have focused on breaking up Facebook, Google, and Amazon. Breakups may be part of the solution, but they are not the entire answer, nor should they be the first step. The initial goals should be to increase competition, to enable alternative business models to flourish, and to protect customers, suppliers, users, and competitors from predatory business practices. Shoshana Zuboff makes an important point that breaking up the platforms without first requiring business model changes will radically increase the number of surveillance capitalists. Smaller, more nimble surveillance capitalists would almost certainly increase the energy behind the business model as each player pursues greater scale. While breaking up Facebook, Google, Amazon, and Microsoft might stall new initiatives like Sidewalk Labs and Libra, which would be a good outcome, it would not likely reduce the harm to public health, democracy, and privacy from existing forms of surveillance capitalism. It might not even improve competition or enable alternative business models, because the various components of today's platforms would still control access to the users that startups would need to succeed.

A better place to start would be antitrust remedies that empower would-be competitors. My goal would be to foster a new industry that treats consumers as customers rather than as products or fuel. If history is any guide, the business opportunity for technology that empowers users would ultimately be larger than the one that currently powers Facebook, Google, Microsoft, and Amazon. The challenge will be to create breathing room to get the industry started. The regulation of the AT&T telecommunications monopoly between 1956 and 1984 provides a useful starting point, as well as a road map. It started with restrictions on AT&T's market opportunity and the free licensing of intellectual property. Subsequent antitrust actions opened up the telecom equipment market to competition and enabled startups in long-distance services to use AT&T's network infrastructure. Only after implementing those changes did the DOJ break up the AT&T monopoly. Applying the AT&T framework can enable a new generation of startups with business models designed to empower the consumers who use their products.

Limiting the platforms' business opportunity is an obvious move, but where to set the limits is less so. For example, forcing Google, Facebook, Microsoft, and Amazon to divest or abandon new initiatives such as smart cities, reserve currencies, transportation, financial services, and smart devices would serve the public interest by limiting the scope of established monopolies and preventing them from taking control of services normally provided by government. Beyond that, where should the lines be drawn? Should Google be forced to divest the infrastructure that supports the online advertising industry? That seems like a very good idea. How about YouTube? Waze? Should Facebook be prevented from going into dating services? Should it divest its online marketplace? Instagram? One or more of its messaging platforms? I favor the tightest limits but recognize the legitimacy of other viewpoints.

Reducing the monopolistic leverage of platforms' intellectual property is another avenue for promoting competition. Internet platforms have huge patent portfolios and cross licenses with each other, both of which block startups. Free licenses would help startups, but in all cases I would favor setting limits on who may benefit. I see no reason to give foreign companies access to the free licenses. Limiting free licenses to US-based, US-owned companies would promote the startup economy, especially if the terms were different for established companies. Perhaps established US companies might pay a royalty for using the platforms' intellectual property that could be used for a good cause, such as compensation to victims of harm. The same idea might work when a licensee sells out to a larger company.

The nature of intellectual property has changed since 1956, necessitating additional measures. The network effects built up by internet platforms are essential to protecting their monopolies. The remedy used against AT&T to promote competition in long distance serves as a useful guide to reducing the power of network effects. Regulators could require that internet platforms provide free advertising to startups to help them acquire users. It would not be crazy to make acquisition of the first one hundred million users free, with small but increasing charges thereafter. Easy access to new customers would reduce or eliminate the need for data portability, avoiding a range of privacy challenges. Again, I would favor limiting this program to startups operating in the United States, owned by US citizens. To qualify for the free license and free advertising, companies must forgo the business model of surveillance capitalism. If they use advertising-based business models, microtargeting would not be allowed.

There is more to antitrust law than enabling competition. The Federal Trade Commission Act and the Clayton Act, both passed in 1914, were designed to protect consumers and markets from monopolistic business practices. Among the obvious regulatory failures relative to internet platforms were Facebook's acquisition of Instagram and Google's acquisition of Waze. Instagram proved to be a powerful competitor to Facebook, dominating the market for those under thirty, but the acquisition enabled Facebook to capture all the benefit. Similarly, Waze might have competed successfully with Google Maps had it remained independent. Instead, both acquisitions cemented monopoly positions. Similar arguments can be made about Microsoft's acquisitions of *Minecraft* and LinkedIn, and Amazon's purchase of Whole Foods. The Clayton Act may provide a legal basis for nullifying each of these acquisitions, but a critical question would remain: What happens to the data shared between the affected products? If every product winds up in possession of all the data currently controlled by the parent, we would be no better off than we are today.

I am not aware of a precedent for regulating the sharing of data sets within internet platforms. Protecting consumers from harm caused by data will almost certainly require a new regulatory vision. It may require new laws and regulations, and perhaps a new agency to enforce them. The internet giants point to Snapchat as an example of why antitrust regulation is not necessary, but the example actually proves my point. A platform with a single data set, like Snapchat, is no match for one with many, no matter how good the single data set is.

At one time, Snapchat enjoyed functional superiority relative to its primary competitor, Facebook's subsidiary Instagram, but has steadily lost ground because of Facebook's multiple levers of monopoly. Facebook's proprietary virtual private network, Onavo, spied on users of Snapchat to see which features were most popular. Leverage from the user base of Facebook allowed Instagram to grow far larger than Snapchat. Facebook's advertising technology enabled Instagram to monetize faster and better. Instagram took advantage of Facebook's advertiser base to lure a larger and more diverse set of advertisers. Game. Set. Match. Snapchat may survive, but it is unlikely to reach the potential it had before Facebook acquired Instagram.

One principle suggested by the Clayton and FTC acts was the notion that corporations should either operate a marketplace or participate in it, but not both. Operators of marketplaces have enormous power that can be abused by favoring their own products or services. Google, Facebook, and Microsoft appear to disregard this principle in the context of their advertising networks, Amazon in its retail operations. Why should Amazon be allowed to use data it captures about the most successful products in its marketplace to create private-label competitors that it favors in its search results? Why should Google be allowed to promote proprietary products in its advertising marketplace? Why should Google be allowed to own many of the core elements of the advertising infrastructure used by the entire internet? Competitors are not the only ones that may be harmed by Google's domination of advertising infrastructure. Marketers who advertise on Google (or other platforms) depend on performance data supplied by services operated by Google. Despite evidence of click fraud and other systemic issues in online advertising, Google's control of a leading measurement service suggests that it will not be addressed without government intervention.

One challenge that government faces in reforming internet platforms relates to the current legal framework for antitrust. Since the early eighties, courts have determined that traditional elements of monopoly behavior—including restraint of trade, predatory pricing, conflicts of interest inside marketplaces—are no longer grounds for antitrust intervention. Thanks to a long campaign by business interests, today's antitrust law employs a single measure of consumer harm: price increases. Internet platforms have created the illusion that they are immune because their services have no monetary cost. That has allowed them to exercise monopoly power over competitors, suppliers, and marketers with impunity. At a breakfast with Barry Lynn of the Open Markets Institute in September 2018, I had an insight that should have been obvious to me sooner: internet platforms are not free; they barter services for data. In such a barter transaction, the currency is data. The true measure of whether prices to consumers have increased would be captured in the change in value of the data supplied relative to the services received. Because platforms already have a lot of data about each user, one might expect the value of incremental data to decline, while the value of the services received would likely be flat to down. However, the average revenue per user for Facebook and Google has been rising rapidly for more than a decade, which suggests the hypothesis about the declining value of incremental data may not reflect reality. If someone could model this effectively, the FTC and the DOJ might have grounds for an antitrust investigation and perhaps a prosecution.

That same day, I took the idea to the chairman of the FTC, Joseph Simons, and to the head of the Antitrust Division of the Justice Department, Makan Delrahim. These were my first meetings with both men, and both were receptive. Neither made a commitment—I did not expect them to—but they encouraged me to pursue the idea and share my progress. Thanks to a serendipitously timed meeting with the president of Yale University, Peter Salovey, I quickly found a kindred spirit with domain expertise. Salovey's career as a social psychologist gave him context for my concerns about internet platforms. He introduced me to Yale's world-class economics department. As it happens, the chair of the department, Dirk Bergemann, was already studying the exact question. He needed access to industry experts to ensure that his model would reflect reality. Tristan Harris, Sandy Parakilas, and I provided that input. Our timing proved to be fortuitous, because a benefactor had just endowed a new center at Yale to apply the social sciences to current problems in society. The center was still in formation, and the antitrust issues raised by internet platforms seemed like a perfect fit. By early 2019, Professor Bergemann and his co-author provided me with a draft of their forthcoming paper demonstrating that the barter system between internet platforms and users has effectively included price increases to consumers. The authors had a key insight: the value of data collected from consumers may come more from its impact on others than from the source. They called this the social value of data. This implied that "own your data" is not an adequate solution; we need regulations to protect the innocent and all of society from the misuse of other people's data. I circulated the draft to the Antitrust Division, the FTC, and members of Congress.

The United States is not the only venue for antitrust. The European Union has its own program, with Google as the primary target. Unfortunately, two multibillion-euro fines appear not to have had a meaningful impact. Google and its investors shrugged them off. It is not clear whether the EU's antitrust enforcers have the leverage necessary to force change. Evidence that they may not came in August 2019, when *The Wall Street Journal* reported that a German court "suspended an order of the German Federal Cartel Office to stop Facebook from combining data it collects about users across its platforms, as well as on third-party apps and websites." The court expressed "serious doubts" that the cartel office had the legal authority to stop Facebook.

The use of antitrust law to limit market opportunities and force divestitures would be valuable for improving competition but would not reduce the damage to public health, democracy, and privacy. For those issues, we need other forms of intervention.

THE SECOND PRONG of my agenda for Washington is to address the flaws in the data economy. Where antitrust law has the ability to improve competition and innovation, policy makers can reduce harm to public health, democracy, and privacy all at once by placing limits on the business model responsible for enabling it, surveillance capitalism.

For far too long, policy makers have deferred to industry with respect to the handling of private data. Despite frequent breaches, they have seldom taken steps to protect consumer welfare. As early as March 1970, *The New York Times* published an article warning about the dangers of digitizing personal data. In 1995, *Wired* magazine published a profile of Equifax, which began life under another name just before the turn of the twentieth century and had a history of mishandling consumer data long before the massive breach in 2017. Remedies such as the Fair Credit Reporting Act of 1970 did not prevent corporations from capturing, storing, transferring, and monetizing the consumer data they touched. The negative consequences of these business practices were held in check for decades by the limitations of computer and network technology, but those limitations evaporated in the early years of this century. The incentives to capture and exploit data increased, as did the scope and intensity of the damage from doing so.

Essentially every action we take in public and online leaves a trail of data for someone to capture. Most of that data eventually gets bought, sold, or traded. The result is that data voodoo dolls exist for each of us, even if we never use internet platforms. The greatest harms from surveillance capitalism increasingly fall on the innocent. The victims of mass murderers in El Paso, Pittsburgh, and Christchurch, New Zealand, did not need to be users of internet platforms. Their killer was, and content on those platforms inspired the killings. Surveillance capitalism created the incentive to amplify the hate speech that inflamed those killers. The platforms made choices; people died. It has been happening for years. Platforms have the power to end the amplification of dangerous speech, but do not exercise it. Instead, they gather data wherever it exists. They amplify whatever content creates the most economic value. They create new technologies to capture data from what used to be private spaces. At present there are no serious limits on the collection, transfer, and commercial exploitation of private data. That has to change.

It is worth asking any politician you meet why they permit opera-tors of email platforms, messaging services, and hosted applications, companies like Google and Microsoft, to scan our documents for in-formation that is valuable to them. We do not permit the US Postal Service or a telephone company to scan the contents of mail or calls. That would be a crime. Why is it possible for some health-care data services and apps to sell personal data about us? The Health Insurance Portability and Accountability Act is supposed to prevent that, but there are loopholes. Why are banks, credit rating services, and credit card processors allowed to sell our private financial information? Why is it okay for cellular carriers to sell location data? Why is web tracking allowed? Why is anyone allowed to gather data from minors? These are questions that deserve serious debate. The business practices developed without controversy, but the situation has changed. The costs to inno-cent people and to society demand attention. Laws can be passed at the federal and state levels. What level of responsibility to society do we expect from corporations? At present, we do not seem to expect any at all. To leave that unchanged would be to invite ever-greater harms.

Civilization has long embraced the concept of sanctuary, a place free from intrusion. Homes and houses of worship have traditionally fulfilled the role, but in practice humans have enjoyed privacy in many other set-tings, including public spaces. This was true until the development of low-cost surveillance technology, especially video cameras. Today, the ex-pectation of privacy has been sacrificed in many public spaces in the name of security. Consumers are voluntarily inviting surveillance devices into the home in the form of smart devices with Amazon's Alexa or Google's Assistant. From a beachhead in the kitchen, such devices often spread to offices and bedrooms. Hospitals and some hotels are putting them in what used to be known as private rooms. Being watched or lis-tened to every moment triggers anxiety. Humans are not wired to be their best selves every moment. They require sanctuary for mental health.

Surveillance capitalism is complex and opaque. The challenge is to explain how it works so that every consumer understands. Scores of people have said to me, "I'm a digital native. I think trading personal data for online services is a good deal. I don't mind that my data is out there. I have nothing to hide." All of these statements can be true and still not justify surveillance capitalism. The vast majority of the data in your voodoo doll got into the hands of internet platforms without your participation or permission. It bears little relation to the services you value. The harm it causes is generally to other people, which means that other people's data can harm you. That is what happened to the victims in El Paso, Christchurch, and so many other places.

Do we want the power of roughly three billion data voodoo dolls to be available to anyone willing to pay for access? Would it not be better to prevent antivaxxers from leveraging Google's predictions about pregnancy to indoctrinate unsuspecting mothers-to-be with their conspiracy theory, placing many people at risk of infectious disease? The same question needs to be asked about climate change denial and white supremacy, both of which are amplified by internet platforms. How about election interference and voter suppression? Internet platforms did not create these ills, but they have magnified them. Is it really acceptable for corporations to profit from the algorithmic amplification of hate speech, disinformation, and conspiracy theories? Do we want to reward corporations for damaging society?

I hope you will join me in a thought experiment. Imagine for a moment that my concerns about surveillance capitalism are valid, that the business model cannot help but undermine public health, democracy, privacy, and competition. What, if anything, would you be willing to do about it? If you knew your actions would help to restore the mental health of loved ones, repair democracy, and give you the ability to make choices without fear, would you give up some of the convenience of internet platforms? Would you be willing to make your voice heard in politics? Given our addiction to internet platforms and their convenience, most people may find politics to be the easier path to making a difference.

In the political realm, I believe there are three paths forward. The ideal—the surest way to stop internet platforms from harming public health, democracy, privacy, and competition—would be to forbid all forms of surveillance capitalism: no data voodoo dolls, no web tracking, no third-party market for data, no scanning of email or documents by service providers, no corporate surveillance in public spaces, homes, or offices. These changes would eliminate microtargeted advertising, rolling back the advertising and marketing businesses about twenty years. This scenario would permit only first-party, intended uses of data. You could share your location and identity with Uber for the purposes of securing transportation, but Uber could not share that data with Google or anyone else, or use it for any purpose other than delivering the ride you ordered. These changes would transform the business of internet platforms, making them much less profitable, but public health, democracy, and privacy are worth it.

The goal of this "no surveillance" plan would be to put a genie back in the bottle. Zero basing the data economy would create the ideal environment for a national discussion of what uses of data should and should not be allowed. Despite mistreatment by internet platforms, marketers are addicted to microtargeting and may well fight efforts to end it, despite the threat to their business from Amazon and potentially others. Smart people have also expressed skepticism about the feasibility of this plan, particularly with respect to enforcement. This is a legitimate concern, because US regulators have a history of failing to match their penalties to the damage inflicted. One possible solution would be to create a new agency tailor-made for this problem. Congress will have to pass new laws. A bill to create such an agency is in the works as I write this. It will pass only if there is a big increase in political pressure from consumers.

Both Google and Facebook argue against regulation on nationalist grounds. They claim to be the country's only defense against Chinese domination of technology. There are good reasons to be worried about competition from China in hardware categories like semiconductors, smartphones, and equipment for 5G networks, but Facebook and Google do not manufacture any products in those categories. We should also take steps to protect the intellectual property of American technology companies and restrict access to American markets by Chinese internet platforms, just as the Chinese have limited access to their market by our platforms. But Facebook and Google are not talking about these issues. Where they compete with China is in behavioral manipulation. Why do we want the best and brightest technologists in America to focus on behavioral manipulation? It is hard to imagine a technology that is more hostile to American values. Would we not be better off if our country's technologists devoted their energy to other, more socially responsible applications of artificial intelligence? By my estimate, more than half of all artificial intelligence engineers and scientists in North America are working for Facebook, Google, Amazon, and Microsoft. The vast majority of these engineers are working on behavioral manipulation or things related to it. People used to joke about how the internet meant devoting the best minds of our generation to cat videos, but this is actually worse, because behavioral manipulation undermines self-determination and the social order.

An alternative path to controlling surveillance capitalism is taxation of revenues from microtargeted advertising, a concept popularized by the Nobel Prize–winning economist Paul Romer. If policy makers cannot ban commerce in third-party data and the use of data voodoo dolls, the next best thing would be a very high level of taxation on microtargeting. For example, a tax of 100 percent would make microtargeting economically unattractive; lower rates would have a lesser impact. Given the scale of the Big Four, the proceeds of taxes on microtargeting could be one hundred billion dollars or more, which makes Romer's proposal attractive to many politicians.

The benefit of targeted taxation would be to change incentives. In an ideal world, the tax rate would be so high that internet platforms would adopt new business models. Unfortunately, there are two obvious issues with taxation as a cure for surveillance capitalism: the definition of microtargeting and the ability of internet platforms to morph around regulations. Advertising does the least damage when constrained to a broadcast model. Platforms will fight that. They will try to define "microtargeting" so that it allows them to do most of what they are doing now. The honest way to do microtargeting would be only in cases when consumers opt in at the advertiser or category level. If you are planning to buy a car or refrigerator, targeted advertising could be attractive, whereas if you do not realize you are pregnant, being targeted by marketers who have been sold "certainty" by an internet platform has the potential to be creepy and even dangerous.

The nature of internet platforms allows them to morph around regulation and taxes far more easily than a business with factories. Facebook reduced the impact of the General Data Protection Regulation in Europe by moving 1.5 billion user files from Ireland to the United States. Platforms will employ similar tactics to avoid taxes. So long as personal data can easily be acquired, corporations will have an incentive to dream up ever more invasive business models to exploit it. In that case, taxation would trigger a game of hide-and-seek, as well as endless litigation about the definition of "microtargeting." All of this makes taxation my second favorite path.

The third path toward reform of surveillance capitalism would be to ban some but not all forms. Politically speaking, the bar should be relatively low to ban service providers from scanning emails, texts, and documents running through their platforms for economically valuable data. This would bring the rules for the internet into alignment with those for telecom companies, postal services, and package delivery companies. Another opportunity would be to tighten up health-care information privacy rules under HIPAA to close the massive loopholes that exist today. The same logic could be applied one category at a time to location data, facial recognition, voice recordings, surveillance around the home, data beacons in public and commercial settings, and the like. The capture, transfer, and exploitation of such data without prior, informed consent seems to violate the most basic of American values, the right to liberty and self-determination. It is not clear to me whether the incremental approach to ending surveillance capitalism will be any easier than doing it all at once. Given the current dysfunction in Congress, it may be harder.

One potential avenue to restrain surveillance capitalism would be through the application of securities law. Professionals in advertising and brand management have expressed to me their suspicion that internet platforms consistently overstate their user count, advertising reach, and duration of advertising views. No one knows for certain, as internet platforms have prevented audits, but in 2016 advertisers succeeded in forcing Facebook to admit to overstating views of video ad metrics. There may be grounds for an investigation of the accounting systems behind digital advertising, many of which are controlled by Google. If it is true that overstatements are material in size and date back many years, violations may amount to fraud and may give rise to criminal liability under US securities law. A criminal investigation—and the possibility of jail time in the event of conviction—would be a game changer. It is not clear whether the SEC will choose to launch such an investigation, but it will certainly not do so without a concerted effort by advertisers. The irony is that Amazon in particular is undermining the foundation of advertising and marketing by reducing brand choice to price and convenience. If marketers are not careful, they will miss the window to protect their business.

IN ADDITION TO ANTITRUST intervention and limits on surveillance capitalism, there is one more job for policy makers in Washington, namely to fix flaws in the internet and platforms that exist separately from surveillance capitalism. These include anonymity, which has been a feature of the internet from the beginning, the safe harbor created by the Communications Decency Act of 1996, and the lack of incentives to prevent mishandling of data.

Contrary to the expectations of the creators of the internet, anonymity has turned out to be a terrible thing, enabling bad actors to find each other, organize, and harm innocent people on a scale never previously imaginable. Facebook's initial requirement that users log in with their official school email ensured a form of authenticated identity that worked well enough at small scale. Facebook's terms of service continue to require authenticated identity, but by leaving enforcement to users, the company enables trolls to have free run of the site. Other platforms have made less effort than Facebook to ensure identity. We know from the experience of online comment sections for newspapers that authenticated identity makes a huge difference in the quality of online discourse, especially when combined with human moderation. Unfortunately, at the scale of internet platforms, anonymity makes human moderation ineffective, even in combination with artificial intelligence. Platforms have two giant incentives to fight authenticated identity. First, it might reduce their user count enough to create issues with Wall Street, advertisers, and perhaps the Securities and Exchange Commission. Second, it would introduce friction, both for users and for the platforms themselves. On the plus side, once implemented, authenticated identity would have huge political benefits for platforms. It would reduce the amount of harmful content while also increasing trust by users and policy makers.

The safe harbor provided to platforms by the Communications Decency Act of 1996—known as Section 230—classifies companies like Facebook and Google as platforms, rather than media companies, protecting them from liability for the actions or content of third parties. This is the law Dan Rose cited when I first reached out to Facebook in October 2016. The passage of Section 230 reflected Congress's understanding that on the internet content moderation would be radically more complex than for a newspaper, magazine, or television network.

Congress acted quickly because by 1996 courts had already begun to penalize companies whose efforts at good faith moderation resulted in lawsuits. Created to give protection to a new industry, Section 230 is still essential for internet startups. Unfortunately, it has become a liability shield for global monopolists against all manner of harms to public health, democracy, and privacy. The argument in favor of Section 230 for all internet players depends on the supposed nonengagement of platforms with respect to content. They claim to be hands off, which may be true in a traditional editorial sense but ignores the deliberate amplification of content that promotes engagement. Platforms create the algorithms and profit from amplification of content like hate speech, disinformation, and conspiracy theories. It is not logical that they do so without any responsibility for the consequences.

For a long time, reform of Section 230 was off the table in Congress. That changed in 2018 when Congress passed the Fight Online Sex Trafficking Act (FOSTA), which created an exception to Section 230 in a narrow category. Several groups are trying to extend the model of FOSTA to categories like terrorism and criminal activities. Senator Josh Hawley of Missouri, who has taken a position of leadership in the Republican Party on issues related to internet platforms, introduced a bill that would alter Section 230 to require internet platforms with more than thirty million users to prove to the FTC that they are free from politically biased moderation. The notion that platforms should not exhibit bias in moderation may sound reasonable but might not be constitutional. If the platforms are not media companies, they are corporations not subject to editorial rules. If they are media companies, then Hawley's bill violates the First Amendment. Tellingly, there is no evidence of political bias in moderation on any internet platform. There are good arguments for reforming Section 230, but the one underlying this bill does not appear to be among them.

I strongly support freedom of expression on internet platforms and everywhere else. I do not want corporations to censor content. My preferred reform of Section 230 would be to remove the safe harbor for algorithmic amplification. At present, platforms are able to tune their algorithms to maximize profits without fear of legal repercussions, even if some of the content they amplify consistently produces harmful outcomes. It is not crazy to suggest that such tuning represents an editorial choice with harms that are predictable in advance. In this case, the safe harbor of Section 230 leads to harm for both individuals and society. Platforms like Facebook, YouTube, and Instagram do not want limits on algorithmic amplification, which may explain their decision to ban people like Alex Jones. For the platforms, it is better to lose all the engaging content of one person than to lose the right to amplify many equivalent voices.

With respect to Section 230, I hope that policy makers will consider reform efforts aimed at choices made by platforms where a reasonable person would anticipate a harmful result. Amplification of antivaxx conspiracy theories or disinformation that might result in violence can have only negative outcomes. Platforms should do everything in their power to limit such content, not promote it. When they fail, there should be liability.

There are similar issues that worry me but may fall outside the scope of Section 230. What should be the penalties for enabling election interference or civil rights violations? Facebook's explanation that it could not have anticipated the election interference of 2016 or the "classic ethnic cleansing" of Myanmar's Rohingya minority in 2017 may be reasonable, but no excuse should be acceptable the second and third times around. Election interference has occurred multiple times since 2016 and remains a threat on Facebook, Instagram, WhatsApp, YouTube, Google, Twitter, and perhaps other platforms, from both domestic and foreign actors. The platforms have paid lip service to addressing it, showing more concern for protecting their business models and privileges than for the integrity of elections at home and abroad. Let me return to my chemical industry analogy. When internet platforms create toxic digital spills, they should pay the cost of the damage they create.

Another issue is data breaches. Yahoo, Marriott, Adult Friend Finder, Equifax, Facebook, Google, and many others have suffered massive breaches. New ones occur with alarming frequency, possibly indicating inadequate incentives to safeguard data. The damage from such breaches can be huge but is not always obvious. The evolution of technology suggests that the scale and damage from such toxic digital spills will increase. We should anticipate breaches of every kind of data. One day there will be breaches of Alexa audio archives and Google's data voodoo dolls. What will be the penalties for Amazon and Google in those cases?

Artificial intelligence is a massive consumer of data, and it is currently unregulated. As a consequence, AI applications are reproducing the worst flaws of the physical world and packaging them in inscrutable black boxes. Some early artificial intelligence applications that manage policing, review résumés, and process mortgage applications have inherited implicit racial and gender bias because the developers did not take due care. (A gloomier possibility is that the buyers of the software wanted to retain implicit bias but hide it in a black box.) What should be the penalty to a software developer for failing to prevent implicit bias in a system that has the potential to cause harm?

When it comes to data, my goal is to place the interests of consumers above all. Let's change the incentives for organizations that hold our data by applying the fiduciary rule to them. Like lawyers and doctors, holders of data should subordinate their own interests in favor of those of the people represented by that data. It should not be legal to hold data without taking effective measures to protect it. The consequences of failure should be far more serious for organizations that suffer breaches than for the people whose data is compromised. The penalties should also exceed any possible benefit to the platforms. This model can also be applied to artificial intelligence applications. Before deployment, every AI application should be required to demonstrate safety, efficacy, and freedom from bias. Preventing implicit bias may be costly, but it should be a requirement in any product where there is no way to audit the decision process, which is currently the case with AI. In addition, all consumers have a right to know which corporations hold data on them, what that data is, to whom it has been transferred, and how it has been exploited. Informing consumers will be expensive, which is a good thing. Raising the cost of exploiting data is in the best interest of consumers.

In the context of regulating internet platforms, the good news is that the issues are systemic, which means they can be addressed with regulation. The bad news is that the issues in question reflect the culture of American industry. The country continues to trust markets more than government, even though markets consistently prioritize shareholders over all other stakeholders. Now that we can see the flaws in the assumption that markets are always the best way to allocate resources, what will we do about it? Can the institutions of government regain their mojo? Can they adapt to the needs of an era dominated by technology? Will we, the people, reinvigorate civic engagement? Restoring trust in government—even if it is only for the purpose of consumer protection—is an essential first step, the foundation for building the movement necessary to effect change.

A LARGE PERCENTAGE OF AMERICANS, quite possibly a majority, now recognizes that a dark side exists for internet platforms. Policy makers in the Trump administration and Congress have taken notice. In comparison to April 2017, when Tristan Harris and I joined forces, the change in awareness and willingness to engage is extraordinary. Facebook, Google, Amazon, and Microsoft are going to fight every action their critics take, as is their right. It won't be a fair fight unless consumers join in. Platforms have all the advantages, not the least of which is their ability to manipulate the attention of and information available to billions. They have few fixed assets and balance sheets laden with cash, as well as the ability to minimize the impact of regulations by adapting around them. In one prominent case—the FTC consent decree of 2011—Facebook was willing to ignore a commitment it made to the government. It took the FTC eight years to impose a penalty, and investors greeted the record five-billion-dollar penalty as a gift to Facebook.

Governments have much less leverage over internet platforms than over traditional businesses, and the platforms know it. In theory, governments could respond to election interference and other issues with a credible threat to shut down platforms, at least temporarily. The government of Sri Lanka did that—and followed through—when the country was beset by online hate speech. Other countries should take note. A credible threat of a temporary shutdown may be the best way to change behavior, but the threat would be credible only if consumers supported it. Consumers in the United States and most other developed countries do not yet view the dark side of internet platforms as a serious enough threat to justify a temporary shutdown, though they might accept one in the event of a terrorist attack. Shutdowns—and the credible threat of them—would be the fastest way to change the behavior of internet platforms, and I have been doing my best to make them acceptable to consumers in more situations.

Nearly three years into this journey, I have concluded that the damage being done by Facebook, Google, Amazon, and Microsoft to public health, democracy, privacy, and competition is greater than the impact of the most extreme acts of terrorism. The platforms have resisted change to their business models, business practices, and culture. We cannot afford to wait for a miracle. We need to force change. A credible threat of temporary shutdowns may be the best way to make that happen. The ball is in our court. The next chapter will provide a road map you can take so that you can be part of the solution.

16

What Each of Us Can Do

The future is already here—it's just not evenly distributed.
—WILLIAM GIBSON

A dystopian technology future overran our lives before we were ready. As a result, we now face issues for which there are no easy answers without much time to act. We embraced smartphones as a body part without understanding there would be a downside. We trusted internet platforms to be benign. We reacted too slowly to warning signs. In the eighties and nineties, the cultural critic Neil Postman warned us that television had ushered in Aldous Huxley's Brave New World. Where Orwell worried about the burning of books, Huxley argued that the greater risk would be citizens no longer wanting to read. Postman predicted that television would entertain us to death. He did not live to see the smartphone prove his argument.

How about you? Are you concerned about children getting addicted to texting or video games? Do you worry about all the kids who cannot put down their smartphones or video games? Have you known a preteen girl who has been body shamed online or a teen who has suffered from fear of missing out? Do you worry about the consequences of foreign countries or other bad actors interfering in our elections? Do you wonder about the moral implications of American products being used to promote ethnic cleansing in other countries? Do you have trepidation about constant surveillance? Are you concerned about being manipulated or denied opportunities by artificial intelligence? These are not hypothetical questions.

Consider your own usage patterns on Facebook and other platforms. What kinds of things do you post? How often? Is any of it inflammatory? Do you try to convince people? Have you joined Facebook Groups dedicated to political issues? Do you get into fights over ideas on social media? Are there people whose posts consistently provoke you? Have you ever blocked people with whom you disagree? Do not feel bad if the answer to any of these questions is yes. The evidence is that most users of Facebook have done these things at one time or another because the algorithms are designed to promote that kind of activity. Now that we know what is going on, what are we going to do about it?

The goal of this chapter is to explain how to protect yourself and your loved ones from the dark side of internet platforms. This journey has transformed my understanding. I see both the problems and the solutions differently than I did when I finished writing the hardcover edition. What each of us does personally still matters, but the structural change we need will only happen if we come together and make our voices heard.

Step one is to understand and accept internet platforms for what they are, rather than what we wish they were. Then I will share what you can do to be part of the solution. There are three areas of opportunity for action: political engagement, personal use of technology, and helping loved ones and friends.

MOST OF US TRUST that internet platforms reflect reality. We trust search results on Google. Most of the time we trust news we read on

Google, Facebook, or Twitter. We assume that the videos on YouTube, the content on Instagram, and the posts from Facebook Groups are authentic. Often they are. But too often, they are not. And there is no easy way to tell the difference between authentic and inauthentic content.

In the current environment of trust, the content of bad actors has far too much power. To reduce the power of bad actors, we need to dial down our trust in the platforms. Danny Rogers, the founder of the cybersecurity firm Terbium, recommends that everyone view Facebook, Google, Instagram, YouTube, Twitter, and the rest as works of fiction, in the way we would a novel, a video game, and a science fiction movie. It is fine to immerse yourself, but you have to remember that the content is not real. It is a simulation. If Facebook or Instagram were a movie, there would be a subtitle, "Based on a True Story."

Thanks to his cybersecurity business, Danny spends time studying online hate speech, disinformation, and conspiracy theories. He has noticed that some conspiracy theories, like QAnon and antivaxx, become a way of life. They have a backstory, a culture, events, rituals, and sometimes a unique vocabulary. The problem is that adherents come to accept that way of life as their own. It becomes their reality. Something similar has happened with the influencer culture on Instagram.

Now imagine how different the world would be if the people "playing" QAnon or Instagram influencer viewed the experience as a game, a simulation, rather than real life. Would either be any more harmful than a video game? There are people who lose themselves in video games, but probably not enough to undermine society in the way internet platforms routinely do today.

It would make a huge difference if the platforms would take the lead on this, educating users that the content on their sites is a fictional representation and should not be taken too seriously. At this point, I am not confident that the platforms will do so.

———

ONCE WE UNDERSTAND THE TRUE nature of internet platforms, the solutions and our opportunity to contribute to them become clear. Believe it or not, the path that requires the least effort from each of us is politics.

We have more power than we realize. Policy makers and internet platforms cannot survive without us. That said, both are happiest when consumers trust them and go along for the ride without paying close attention. Where politicians differ from internet platforms is in their response to public pressure. Politicians respond to voters. If enough of us make our voices heard about the need for reform of internet platforms, then policy makers will help us force change. That is our opportunity.

With the 2020 US presidential election fast approaching, we can make our voices heard. What makes this issue so unusual is that it cuts across the political spectrum. My activism has introduced me to the widest range of audiences, but no matter their politics on other issues everyone reacts more or less the same way to my message. Audience members leave the room with an understanding of the problem and a sense of urgency to do something about it. The United States may be polarized on everything else, but the issues surrounding internet platforms are an oasis of harmony. This situation is not lost on Republican and Democratic members of Congress, or on members of the Trump administration. On this issue, bipartisan solutions may be possible . . . if enough of us get involved.

What does it mean to "get involved"? Most of us do not engage directly in politics and may not realize how easy it is to do so. When I began my activism in 2017, I knew no one in Washington but met dozens of members of Congress in the first year. It turns out that every politician wants to hear from constituents. You can start by sending an email or making a phone call to your member of Congress and your senators. There are email forms and phone numbers on every congressional website that can be found with a web search. You can meet your member of Congress or staff members in their district office or in Washington. All you have to do is make an appointment. House members face an election every two years, so most spend lots of time in the district doing town halls and events. It is also useful to share your views with mayors, governors, state attorneys general, and members of your state legislature. Every one of these people can play a role in the solution. Something that may surprise you: meeting politicians is fun. Give it a try.

What should be your message to politicians? If there is an aspect of the problem that appeals to you—public health, kids, democracy, privacy, competition, or innovation—go with it. You will be most effective on issues you care about deeply. Policy makers increasingly understand the overlapping nature of the harms from internet platforms, so stick with the issues that concern you most. Presumably, they will be the ones you understand best.

When I have a politician's attention for only a short time, I focus on surveillance capitalism. My first question is, "Why is it legal for internet platforms like Google and Microsoft to scan my email, messages, and documents for information that is valuable to them? Why aren't emails and online documents treated like phone calls and letters?" From there, I go to the other problematic forms of surveillance. "Why doesn't the health-care data privacy law [HIPAA] prevent all corporations from monetizing my most personal health data? Why is it legal for financial institutions, cellular carriers, app vendors, and other corporations to transfer and exploit my private data for profit? Why is any corporation allowed to track me online? Why are corporations allowed to collect data about children under 18? Why are the vendors of smart devices with Alexa and similar software permitted to spy on us and then exploit the data?" If you are concerned about climate change, gun violence, or the antivaxx conspiracy theory, make certain that your elected representatives understand the role that internet platforms play in preventing solutions. Politicians will figure it out quickly. Scanning email and exploiting health data happen because of legal loopholes that should not exist and that Congress can close immediately. States could do it, too.

Thanks to support in Congress and in state legislatures, there is huge value in focusing your energy on the mission of protecting children from internet platforms. Studies of the country's experiment with exposing children to technology early have raised alarms. The evidence suggests that delaying exposure and minimizing it for as long as possible would be better for kids. Unless there is no alternative, do not babysit or entertain young children with phones or tablets. The problem goes beyond excessive dopamine stimulation, which can undermine the ability to concentrate on a permanent basis. The content itself can be harmful. For example, YouTube Kids is plagued with age-inappropriate content. Congress should regulate that. Ask your elected representatives why it is legal for corporations to gather, transfer, and exploit data about children under eighteen. Due to their widespread use in schools, Google Chromebooks and Classroom are the entry level of surveillance capitalism. The argument that the data is anonymous is disingenuous. The systems have so much data on each child that determining identity is not a technical challenge. Minors' civil rights are being violated at an age when they are not allowed to give consent.

Another issue that Congress needs to address is the age at which children may be targeted by internet platforms. Thanks to the Children's Online Privacy Protection Act, which became effective in 2000, children under the age of thirteen get special protections. Above that age, children are legally considered fair game. The online world has changed dramatically since 2000, and there are strong arguments for extending the protections of COPPA to children between thirteen and eighteen and strengthening the protections for all children. Given its small size, the FTC's recent fine against YouTube for violating COPPA seems unlikely to change behavior.

Thanks to Cambridge Analytica, privacy protections are a hot topic of conversation in the halls of government, and not just for children. The early efforts—the General Data Protection Regulation in Europe and the California Computer Privacy Act in this country—are well intended but do not reflect the breadth of the problem as we understand it now. The first political priority—which can be implemented in Congress or in states—should be to give all users of internet platforms the right to sue for damages if they believe they have been harmed. The terms of service of every platform require arbitration in the event of conflict, a situation that generally favors platforms. Given what we know now, that is not reasonable. This is a straightforward case to make to any politician.

The right to sue is low-hanging fruit, but the big privacy prize would be to end surveillance capitalism. In Washington, there is no bill yet to prohibit the use of data voodoo dolls, but some members of the House of Representatives are committed to the next best thing, data minimization. The idea is to prevent the transfer or third-party monetization of personal data. First-party intended use of data would be fine. Third-party use would not. Companies would be able to use your data only to provide the service you are securing at the time. No other use or transfer would be permitted. By resetting privacy rules this way, we would create a context for negotiating uses of data that would give consumers as much control and power as possible.

If ending surveillance capitalism is not possible, the next best thing would be to give each person control over his or her own data. Congress needs to pass legislation that requires corporations that hold data to inform consumers about the data they hold, where it came from, where it has gone, and how the holder has monetized it. Some advocate for giving consumers ownership over their data and the ability to monetize it. That would certainly be better than what we have now, but does not take into account the consequences to third parties. Personal data should be treated as a human right, not an asset. I share Shoshana Zuboff's view that personal data—and data voodoo dolls—are like human organs; they should not be bought and sold. If you meet politicians running for office, I hope you will tell them to protect privacy by ensuring that consumers know who holds all their data and how it is being used . . . and have complete control of how it is used. This may be another thing that can be done at the state level.

Protecting elections is a mission distributed among the federal government, states, and counties, which is one reason why the infra-structure is vulnerable to hackers. Treating election infrastructure as strategic infrastructure—equivalent to the electrical grid—should be a priority, as should be federal control of federal elections. The status quo enables bad actors at the state and county levels—including elected officials—to use election infrastructure for partisan gain.

Another important action Congress might take would be to prohibit microtargeted advertising in the context of elections. Microtargeting has transformed marketing by enabling the advertiser to pay only for the au-dience it desires. It works incredibly well in politics, so it is hard to imag-ine that Congress will pass legislation in the absence of overwhelming pressure from voters. Platforms may also raise First Amendment objec-tions. Congress may be more willing to create federal penalties for en-abling election interference. Most state and local governments do not have the expertise to deal with election interference. Each of us has an opportunity to persuade them to take the issue more seriously.

The dark side of social media includes other problems that can be eased or eliminated through government intervention. Examples include anti-trust, the fiduciary rule, and identity, all of which I covered in chapter 15.

PROTECTING OURSELVES FROM the dark side of internet platforms begins with recognition that the thing that makes internet platforms so compel-ling, their convenience, is actually a terrible thing for users. Convenience is a drug. We cannot get enough of it. We seek it even when we know there may be harmful consequences. Products that bring convenience to daily life became a big business after World War II, evolving into the mass cus-tomization embodied in Burger King's motto, "Have it your way." In the early days of the twenty-first century, Google and then Facebook did for ideas what consumer package goods had done for products, which brought us to our current predicament, billions of *Truman Shows*.

From the start, internet platforms made convenience the focus of their designs, triggering a virtuous cycle of adoption. For consumers who already use Gmail, Google Maps, and search, the benefits of interoperability make adoption of new Google products a no-brainer. Friction disappeared, along with critical thinking about consequences. We crave more. The more apps we adopt from a single vendor, the greater the perceived convenience and the greater the ability of that platform to manipulate our choices and behavior.

By prioritizing convenience, we increase our vulnerability. Friction in our lives allows for adaptation and self-determination. Having run the experiment of embracing friction for the past two years, I have discovered that the convenience of technology was not as helpful as I had thought, and it caused me to miss many wonderful things. I communicated electronically instead of meeting face-to-face. When I did meet people, I let notifications interrupt the conversation. Returning to a life of meaningful interactions, ones with eye contact, has been immensely satisfying. Technology still has a place in my life, but it no longer mediates and undermines important personal relationships.

I was able to reject convenience because I noticed a correlation with potential harm. The correlation is not perfect in every case, but created an incentive to evaluate every app on my phone and computer, every card in my wallet, every service I used. I made many changes.

ONE WAY WE CAN defend ourselves is to change the way we use technology. There is evidence that internet platforms produce happiness for the first ten minutes or so of use, but beyond that continued use leads to progressively greater dissatisfaction. The persuasive technologies embedded in the platforms keep users engaged. We cannot help but scroll down just a bit farther, in the hope of something really wonderful. We did not understand the dark side of internet platforms before we got hooked, but we can modify our behavior now.

Changing behavior starts with reconsidering your relationship to the internet platforms. That is what I did, especially for Facebook and Google. (I have never used LinkedIn; the product's exploitation of user address books signaled to me a willingness to invade privacy.) I have given up Google, which I will describe in a moment. I still use Facebook and Twitter but have changed my behavior, especially on Facebook. I no longer allow Facebook to press my emotional buttons. I wish it were not necessary, but I do not post anything political or react to any political posts. It took six months, but my feed is now dominated by the music side of my life, birthdays, and puppies. In addition, I erased most of my Facebook History. To make these changes, I forced myself to pause before I did anything and ask, "What could go wrong?" If I can think of anything at all, I stop right there. Facebook's algorithms amplify disinformation and fake news, so I am really careful about the sources of information on Facebook and elsewhere on the web. To protect my privacy, I do not use Facebook Connect to log in to other sites or press Like buttons I find around the web. I do not use Messenger or WhatsApp. I use Instagram only to promote this book, because there is no other way to reach people who use Instagram.

I avoid using Google whenever possible because of its data collection policies. Avoiding Google is inconvenient, so I have turned it into a game. I use DuckDuckGo as my mobile browser and search engine because it does not collect search data. I use Signal for texting. I do not use Gmail, Google Docs, Waze, or Google Maps. Instead, I use my own email server, Microsoft Office apps for Mac, and Apple Maps. I employ several tracking blockers, including Ghostery and Disconnect, to make it much harder for internet platforms and others to collect data about me. I use Apple Pay whenever I can, because it is the closest thing to cash in the world of online payments. I look forward to trying the Apple Card credit card, which reduces data leakage from payments. I use cash for all transactions up to two hundred dollars for the same reason. I am far from invisible on the web, but my shadow is smaller.

I am very careful about the devices I use. I favor Apple products because that company respects the privacy of my data, while Android does not. The difference between the two platforms is much bigger than Android users realize. Apple has made privacy a strategic priority, while Android has done the same with surveillance. I still check my phone way too often, but I have turned off notifications for practically everything. I allow notifications only for texts and baseball, and for them I prefer vibrations. I read books on my iPad, so I keep it in Night Shift mode, which takes all the blues out of the display, reducing eye strain and making it much easier to fall asleep at night. I sometimes put my iPhone in monochrome mode to reduce the visual intensity of the device and, therefore, the dopamine hit. I do not recharge my devices in the bedroom. A piece of duct tape covers the camera on my laptop.

I bought an Amazon Alexa smart speaker on the first day it was available. About an hour after I installed the device in the kitchen, an ad for Alexa came on the TV, and my Alexa responded. I realized immediately that Alexa would always be listening and that no one should trust it to respect privacy. My brand-new Alexa went into a storage container, never to return. Unfortunately, the entire world of connected televisions, appliances, and other devices—the Internet of Things—shares Alexa's snoopiness. Unscrupulous vendors can use IoT devices for surveillance. Incompetent vendors may leave customers vulnerable to hacking by bad actors. All vendors will collect masses of data; no one knows what they will do with it. My advice is to avoid IoT devices, or at least avoid putting them on your network, until vendors commit to strong privacy protection.

Data privacy is sufficiently abstract that few people take steps to protect it. I recommend using a password manager such as 1Password to ensure secure access to websites. I am hopeful about Sign In with Apple. It will not initially replace a password manager, but it should be a big step forward relative to competing products from Facebook and

Google. My advice is to approach every new product—and all the old ones you already use—with caution until you confirm each product is benign. The internet platforms and IoT hardware companies have yet to demonstrate a commitment to safeguarding users; it makes no sense to trust them with your data or anything else.

IF YOU ARE A PARENT of a child eighteen or younger, think about whether you are a good digital role model. How often do you check devices in the presence of your kids? Monitor your usage for a few days to see. How much do you know about your children's online activity? What is the ratio of online activity to outdoor play in your family? How often do you go online together with your kids? To what degree does your children's school employ computers and tablets? At what age did that start? How often do your children's classes engage in traditional group learning, where the students are encouraged to participate with classmates, as opposed to one-on-one activity?

For parents who want to understand the impact of persuasive technology on children, I recommend *Irresistible: The Rise of Addictive Technology and the Business of Keeping Us Hooked* by New York University professor Adam Alter and *Glow Kids: How Screen Addiction Is Hijacking Our Kids—and How to Break the Trance* by Nicholas Kardaras. Online resources include the websites of the Center for Humane Technology and Common Sense Media.

The medical profession has a message for parents: smartphones, tablets, and the apps that run on them are not good for children. When it comes to protecting children, there is mounting evidence that the only thing kids should do on smartphones is make calls. Just about everything else kids do on smartphones poses a threat of one kind or another. We used to think that kids were safer playing with technology than playing outdoors. We were wrong.

At the margin, research suggests that less exposure to devices and apps is better at all ages but essential for younger children. The American Academy of Pediatrics suggests that parents not allow any screen exposure for children under two. There is a growing body of research indicating that parents should limit the screen time of kids under twelve to much less than current averages. The rationale for early exposure—preparing kids for life in a digital world—has been negated by evidence that screen time impedes childhood development to a far greater degree.

Parents, teachers, and PTAs should challenge the presence of devices like Google Chromebooks in elementary school. Pediatricians have told me that classrooms should be where children learn to focus on the teacher and socialize with classmates. Screens interfere with both, even if the content is academic. Given how deeply smartphones, tablets, and PCs have penetrated everyday life, cutting screen time is much easier said than done. Many schools insist on using PCs or tablets in the classroom, despite evidence that computers and tablets may be counterproductive in that setting. It may not take much screen time to trigger an excess of dopamine in kids. Doctors have observed that kids with too much screen time suffer from a variety of developmental issues, including an inability to pay attention and depression. With preteens and teens, overproduction of dopamine remains an issue, but the exploitation of social media by bullies is also a problem. Even kids in their teens are unprepared to cope with the addictive powers of internet platforms on mobile devices. The success of YouTube with young people also presents a challenge: kids assume the videos they watch are accurate, which makes them vulnerable to conspiracy theories and other nonsense. The Google Classroom applications suite is pervasive in schools, enabling Google to capture massive amounts of data on kids. Even if the data were truly anonymous, which may not be the case, it would provide Google with powerful insights about the behavior of children too young to give consent.

Parents face a daunting challenge. Even if they control technology at home, how do they protect their children elsewhere? There are devices everywhere, and too many people willing to share them, un-aware of the danger. One possible first step is to organize small groups of parents in a format equivalent to a book club. The goal would be to share ideas, organize device-free playdates, and provide mutual support toward a goal that requires collective action. Small groups can be an effective first step in a long campaign to change the cultural zeitgeist.

Adults are less vulnerable to technology than kids, but not enough so as to be safe. Smartphones and internet platforms are designed to grab your attention and hold it. Facebook, YouTube, and other platforms are filled with conspiracy theories, disinformation, and fake news disguised as fact. Facebook and YouTube profit from outrage, and their algorithms are good at promoting it. Even if the outrage triggers do not work on you, they work on tens of millions of people whose actions affect you. Elections are an example. Facebook in particular has profited from giving each user her or his own reality, which contributes to political polarization. Facebook has managed to connect 2.4 billion people and drive them apart at the same time.

I would like to think that the people who make smartphones and internet platforms will devote themselves to eliminating the harmful aspects. Apple has taken meaningful steps to protect privacy, but it has a lot of hard work to do if it wants to reduce addiction. Android's privacy profile is much worse than Apple's iOS, and its addiction profile is comparable, which is to say it is terrible. Policy makers in Washington and the states can create incentives for the internet giants to do right by their users. No one else has a big enough megaphone to get the platforms' attention.

The platforms act as though they believe users will be too preoccupied to look out for their own self-interest, to insist on dramatic change. Let's prove them wrong. Parents, users, and concerned citizens can make their voices heard. We live at a time when citizens are coming together to bring about change relative to civil rights, labor rights, and gun violence. Let's do the same about reining in internet platforms. Bringing people together in the real world would be the perfect remedy for addiction to internet platforms. If we can do that, the world will be a better place.

Epilogue

We have to take our democracy back. We cannot leave it to Facebook or
Snapchat or anyone else. We have to take democracy back and renew it.
Society is about people and not technology. —MARGRETHE VESTAGER

Freedom is a fragile thing and never more than one generation away
from extinction. —RONALD REAGAN

Four years have passed since I first observed bad actors exploiting
Facebook's algorithms and business model to harm innocent people. I did not imagine then the damage to democracy, public health,
privacy, and competition that would be enabled by internet platforms I
loved to use. If you live in the United States, the United Kingdom, or
Brazil, your country's politics have been transformed in ways that may
persist for generations. In Myanmar, Sri Lanka, New Zealand, and the
United States, your life might have been threatened. In every country
with internet access, platforms have transformed society for the worse.
We are running an uncontrolled evolutionary experiment, and the results so far are terrifying.

My goal in writing *Zucked* was to provide readers with the information they would need to participate in a national conversation about the dark side of internet platforms. Thanks to Shoshana Zuboff and others, my understanding of the problems is more nuanced and complete than when I finished writing the hardcover edition. There will always be more to learn, but we know enough to act. We cannot use "we need to study this more" as an excuse for inaction.

As people, and as citizens, we were not prepared for the social turmoil and political tumult unleashed by internet platforms. They emerged so quickly, and their influence over both people and commerce spread so rapidly, that they overwhelmed cultural, political, and legal institutions. And yet, too many of us continue to trust internet platforms and the content we find on them. Some will be tempted to relax now that the 2018 midterm elections have come and gone without obvious foreign interference. Conscientious citizens may stop worrying, comfortable in the knowledge that policy makers are now on the case. That will not work. Instead, I hope everyone will see that foreign meddling in campaigns is merely one symptom of a much larger problem, a problem for which the internet platforms themselves—and nobody else, in or out of government—must be called to account.

In his brilliant book *The Road to Unfreedom*, Yale professor Timothy Snyder makes a convincing case that the world is sleepwalking into an authoritarian age. Having forgotten the lessons of the twentieth century, liberal democracies and emerging countries alike are surrendering to autocratic appeals to fear and anger. Facebook, Google, and Twitter did not cause the current transformation of politics, but they have enabled it, sped it up, and ensured that it would reach every corner of the globe simultaneously. Design choices they made in the pursuit of global influence and massive profits have undermined democracy and civil rights. Snyder's argument dovetails with that of Evgeny Morosov, who believes that the institutions of democracy are not designed for an era dominated by technology and must be rebuilt from the ground up to fulfill their mission of protecting citizens.

Let me be clear. I do not believe the employees at Google, Facebook, or Twitter ever imagined their products would harm public health, democracy, and privacy in the United States or anywhere else. But the systems they built are doing just that. The new systems they are building will do it more effectively and, therefore, more dangerously. With grand plans for smart cities and reserve currencies, Google and Facebook may cross a line from being enablers of harm to being perpetrators. The single-minded pursuit of growth by corporations that do not anticipate unintended consequences and do not believe they should be held accountable for them will always produce undesirable side effects. At the scale of Facebook and Google, those side effects can undermine the foundations of society. It is up to society to protect itself. That means we, as citizens, must engage in the debate and insist on reform of internet platforms. They should not be allowed to usurp the role of government. The unintended consequences of privatization of prisons and services provided to the military have produced great harm. We should not count on better outcomes with respect to smart cities and reserve currencies sponsored by internet platforms.

I believe a strong case can be made that Facebook and Google are threats to national security. Businesses that undermine public health, democracy, and privacy pose a clear and present danger. They should not be free to operate without supervision. Platforms that allow other countries to interfere in our elections should trigger outrage and regulatory intervention. Google, which claims AI is a strategic technology, has been so eager to do business in China that it volunteered to host its most advanced artificial intelligence in a facility vulnerable to intellectual property theft. The story sparked enough outrage from Congress that Google withdrew the proposal, at least temporarily.

When I wrote the hardcover edition, I thought the most effective way to force change would be for users to take the lead, mostly by changing their online behavior. Such changes are still important, but I now believe the best path to reform is for consumers to exercise their political power, in partnership with policy makers. Internet platforms have demonstrated the ability to fend off government and consumers when they act alone. My hope is that by teaming up, consumers and policy makers will have more leverage. This will require a cultural change, but there are signs that the country may be ready for such a change.

In a perfect world, users would escape filter bubbles, not allow technology to mediate their relationships, and commit to active citizenship. Growing numbers of Americans are doing precisely that. Civil rights movements like Black Lives Matter, the Women's March, and March for Our Lives; labor actions by teachers, air traffic controllers, and the workers at Stop & Shop; and massive demonstrations by the residents of Puerto Rico are encouraging signs. Turnout in the midterm elections is another. But the midterms also demonstrated that democracy cannot be restored in a single election cycle. A large minority of Americans remain comfortable in an alternative reality enabled by internet platforms.

The platforms want us to believe that surveillance capitalism is necessary for them to deliver the services we love. That is nonsense. What is at risk for consumers from the changes I propose is some of the convenience that comes from entrusting the entirety of your digital life to surveillance capitalists. We need to view internet platforms for what they are: profit-seeking businesses that are happy to sacrifice our well-being to improve their bottom line. We need to view products like Facebook, Instagram, Google, YouTube, Twitter, Reddit, 4chan, and 8chan as forms of entertainment loosely based on real life. We cannot afford to let them rule our lives for the sake of convenience.

I have run the experiment of giving up convenience. It has been more rewarding than I imagined. The convenience forgone has been more than offset by richer personal relationships and a deep sense of satisfaction that I am reasserting my right to self-determination. Your mileage may vary, but there is real value in running the experiment.

Government intervention against surveillance capitalism and monopolistic behavior will enable alternative business models and applications to emerge. Done right, new regulations will reduce the profitability of surveillance capitalists while giving consumers both more and safer choices. Facebook, Google, Amazon, and Microsoft will not go away because of competition. Ideally, they will be forced to share the opportunity with many others and to respect the rights of consumers. The country has faced similar challenges with the robber barons at the turn of the twentieth century, AT&T in the fifties, and the chemical industry in the seventies and eighties. Those experiences do not cover all the issues we face with internet platforms, but they enable agencies like the FTC and the Antitrust Division of the DOJ to get started.

I have spent four years on this journey, from a time when the issues described in this book were on the radar of only a handful of people. Look how far we have come! Who knows where the journey will lead, but I hope you will join me. Bring your family and friends. Together, we can accomplish something that makes us proud. We have many tools available to us. Let's use them all. Let's insist that policy makers do the same. We do not have much time, but we have the numbers. The vast majority of all human beings have a common interest in reducing the power of and harm from internet platforms. We have every right to insist that technology companies stop treating us as fuel and return to making products that empower us.

In my dreams, billions of consumers will rebel, changing the way they view internet platforms and how they use them but also joining with policy makers to force much-needed reform. Perhaps this book will help. In real life, policy makers around the globe must accept responsibility for protecting their constituents. The stakes could not be higher.

ACKNOWLEDGMENTS

I wrote this book to further the cause for which Tristan Harris and I came together in April 2017. Since then, dozens of people have volunteered their time, energy, and reputation to bring the threat from internet platforms to public awareness. It started with my incredible wife, Ann, who encouraged me to reach out to Zuck and Sheryl before the 2016 election and supported the movement at every turn.

I want to thank Rana Foroohar of the *Financial Times* for suggesting that I write a book and for introducing me to my wonderful agent, Andrew Wylie. Andrew in turn introduced me to Ann Godoff and my editor, Scott Moyers, at Penguin Random House. The Penguin team has been amazing, especially Sarah Hutson, Matt Boyd, Elisabeth Calamari, Caitlin O'Shaughnessy, Christopher Richards, and Mia Council.

Many people offered valuable feedback on this manuscript. My appreciation goes to Ann McNamee, Judy Estrin, Carol Weston, Joanne Lipman, Barry Lynn, Gilad Edelman, Tristan Harris, Renée DiResta, Sandy Parakilas, Chris Kelly, Jon Lazarus, Kevin Delaney, Rana Foroohar, Diane Steinberg, Lizza Dwoskin, Andrew Shapiro, Franklin Foer, Jerry Jones, and James Jacoby.

Let me extend thanks and appreciation to the team that created the Center for Humane Technology: Tristan Harris, Renée DiResta, Lynn Fox, Sandy Parakilas, Tavis McGinn, Randima Fernando, Aza Raskin, Max Stossel, Guillaume Chaslot, Cathy O'Neil, Cailleach Dé Weingart-Ryan, Pam Miller, and Sam Perry. Your energy and commitment never cease to inspire me.

Special thanks to Judy Estrin, who, for the better part of a year, met weekly as a thought partner and helped organize the ideas at the foundation of this book. Buck's of Woodside provided the setting for every one of those meetings, once again playing its role as a place where ideas become real. Thank you to Jamis MacNiven, proprietor.

Shoshana Zuboff's *The Age of Surveillance Capitalism* transformed the discussion about internet platforms. Through her book and subsequent conversations, Shoshana guided me to a deep understanding of surveillance capitalism and the business practices of Google and Facebook, informing my activism and the revisions to this paperback edition of *Zucked*. Thank you.

Sheepish thanks go to two dear friends—Paul Kranhold and Lindsay Andrews—who fielded dozens of press contacts without compensation in late 2017 and early 2018 because I had forgotten to take their contact information off the Elevation website when the fund completed its successful journey.

Jonathan Taplin wrote the book that helped me understand the economic and political aspects of what I had seen in 2016. Andy Bast of *60 Minutes* created the original interview with Tristan Harris. Brenda Rippee did Tristan's makeup for *60 Minutes* and *Real Time with Bill Maher*, introduced Tristan to Arianna Huffington, and then did my makeup for *Real Time with Bill Maher*. *Bloomberg Tech*, normally hosted by Emily Chang, brought Tristan and me together. Caroline Hyde, Candy Cheng, Arianna Huffington, Ben Wizner of the American Civil Liberties Union, and US Senate staffer Rafi Martina supported Tristan and me in the early days. Thank you. Chris Anderson and Cyndi Stivers of TED gave Tristan a platform and encouraged us. Eli Pariser provided both inspiration and material aid when we needed it. John Borthwick devoted many hours to helping Tristan and me explore the implications of our hypotheses.

Early on, two organizations committed themselves to our cause. Barry Lynn, Sarah Miller, Matt Stoller, and the team at the Open Markets Institute (OMI) guided us through the halls of Congress in the summer and fall of 2017, making many key introductions. Thanks also to the extended OMI family, especially Tim Wu, Franklin Foer, Zephyr Teachout, and Senator Al Franken. My old friend Jim Steyer founded Common Sense Media to help parents and children manage their media consumption, which led him to be an early critic of internet platforms. Jim and his colleagues provided the initial home to the Center for Humane Technology and organized the events that put it on the map. Common Sense also took the lead on privacy and anti-bot legislation in the California legislature. Bruce Reed's guidance helped us succeed in Washington. Thanks also to Ellen Pack, Elizabeth Galicia, Lisa Cohen, Liz Hegarty, Colby Zintl, Jeff Gabriel, Tessa Lim, Liz Klein, Ariel Fox Johnson, and Jad Dunning.

Senator Mark Warner was the first member of Congress to act on our concerns. We cannot thank him enough for his leadership in protecting democracy. Senator Elizabeth Warren was the first to support traditional antimonopoly regulation of the technology industry. Federal Trade Commissioner Terrell McSweeny helped us understand the FTC's regulatory role.

My first public words about my Facebook concerns appeared in *USA Today*, thanks to a commission from the editor-in-chief, Joanne Lipman. *USA Today* published three of my op-eds over four months, helping me find my voice. CNBC's *Squawk Alley*, where I am a regular contributor, reported on the first *USA Today* op-ed and has vigorously followed the story ever since. To Ben Thompson, Carl Quintanilla, Jon Fortt, Morgan Brennan, and the rest of the *Squawk Alley* team, thank you.

I first met Ali Velshi on the floor of the New York Stock Exchange in May 2017, when I was coming off an appearance on *Squawk Alley*. This was on the first trip that Tristan and I took to New York, and Ali was interested in what we thought was going on. Soon thereafter, Ali got his own show on MSNBC and started covering our story in October. He understands the issues, and his viewers benefit. Lily Corvo, Sarah Suminski, and the amazing makeup and production crews at MSNBC never cease to amaze. I gained insights every time I visited the green room at 30 Rock, thanks to an endless supply of smart, well-informed people. Thank you to Ari Melber, who brought his legal mind to the Facebook story, and to Lawrence O'Donnell for analyzing the story through the lens of his deep understanding of history, politics, and government. Katy Tur is an amazing journalist and was the first target of Trump's attacks on the free press. She is also a serious fan of Phish, which is a plus in my book. Other guests on MSNBC shared important insights. In particular, I want to thank Clint Watts, Joyce Vance, and Eddie Glaude Jr. Thank you also to Chris Jansing, David Gura, Yasmin Vossoughian, Joanne Denyeau, Justin Oliver, and Michael Weiss.

This book would never have happened had it not been for Paul Glastris and Gilad Edelman, who commissioned an essay for the *Washington Monthly*, guided me through the process of writing it, and helped me realize that I was ready to write a book. That essay had far more impact than any of us imagined. Thank you, my friends.

Giant thanks go out to Jeff Orlowski, Lissa Rhodes, Laurie David, and Heather Reisman, who are creating a documentary about addiction to internet platforms and its consequences. My deep appreciation also goes to James Jacoby, Anya Bourg, Raney Aronson, Dana Priest, and the rest of the amazing team at *Frontline* for documenting the impact of social media on the 2016 election. Their film, *The Facebook Dilemma*, won a Peabody Award. In addition, I want to thank Geralyn White Dreyfous, Karim Armer, and their documentary team for digging deep into the Russian election interference. *The Great Hack* documents the Cambridge Analytica scandal as it was unfolding.

Lucky accidents brought our team into the orbit of *The Guardian* and *The Washington Post*. In partnership with *The Observer* in the UK, *The Guardian* broke the Cambridge Analytica story. We worked with Paul Lewis, Olivia Solon, Julia Carrie Wong, and Amana Fontanella-Khan. Thank you all! At *The Washington Post*, we worked most closely with Elizabeth Dwoskin and Ruth Marcus. Thank you both! Thanks also to Betsy Morris for a wonderful profile in *The Wall Street Journal*. Thank you to Brian Zittel and *The New York Times* for the opportunity to contribute an essay to the Privacy Project.

On *CBS This Morning*, Gayle King, Norah O'Donnell, and John Dickerson jumped on the story and told it well to a huge audience. Thank you. Thanks also to Chitra Wadhwani. The team of Jo Ling Kent and Chiara Sottile broke important stories about Facebook on NBC. Thank you. Huge thanks to my longtime friend Maria Bartiromo and her booker Eric Spinato on Fox Business. Tucker Carlson on Fox zeroed in on the threat to kids and teens from social media. On *Fox & Friends*, Steve Doocy focused on privacy. Many thanks to Alexander McCaskill and Andrew Murray. Thank you to Hari Sreenivasan and the team at *PBS NewsHour* for giving me an opportunity to share the story with their audience. ITN's Channel 4 in the United Kingdom broke giant stories about Cambridge Analytica and Facebook. Thank you to all the teams who produced those stories.

Huge thanks to all the radio programs that dug into the Facebook story. In particular, I want to thank CBC Radio in Canada, BBC Radio in the UK, NPR's *Morning Edition*, NPR's *Weekend Edition*, and Bloomberg Radio. Thank you also to Sarah Frier and Selina Wang at Bloomberg.

I tip my hat to David Kirkpatrick, whose interview of Zuck immediately after the 2016 election produced the "it's crazy" response to the question of whether Facebook influenced the vote, for devoting a large portion of his conference a year later to the problems with internet platforms. David's encouragement has meant the world to me.

Thank you to the many journalists and opinion writers whose work has informed my own. Special thanks to Zeynep Tufekci, Kara Swisher, Donie O'Sullivan, Charlie Wartzel, Casey Newton, April Glaser, Will Oremus, Franklin Foer, Tim Wu, Noam Cohen, Farhad Manjoo, Matt Rosenberg, Nick Bilton, Kurt Wagner, Dan Frommer, Julia Ioffe, Betsy Woodruff, Charles Pierce, Josh Marshall, Ben Smith, Brittany Kaiser, Niall Ferguson, Norm Eisen, and Fred Wertheimer (who also guided my approach to Washington).

On Capitol Hill, my appreciation goes out to every member and staffer who took a meeting with me. Thank you to Senators Richard Blumenthal, Cory Booker, Sherrod Brown, Al Franken, Orrin Hatch, Doug Jones, Tim Kaine, John Kennedy, Amy Klobuchar, Edward Markey, Jeff Merkley, Gary Peters, Tina Smith, Jon Tester, Mark Warner, and Elizabeth Warren. Thanks to staff members Elizabeth Falcone, Joel Kelsey, Sam Simon, Collin Anderson, Eric Feldman, Caitlyn Stephenson, Leslie Hylton, Bakari Middleton, Jeff Long, Joseph Wender, Stephanie Akpa, Brittany Sadler, Lauren Oppenheimer, and Laura Updegrove, among many others. Thank you also to the wonderful Diane Blagman.

Representatives Nancy Pelosi and Adam Schiff dug into the issues and have helped our team share its message broadly in the House. Thank you both! Thank you also to Representatives Kathy Castor, David Cicilline, Mike Doyle, Anna Eshoo, Brian Fitzgerald, Josh Gottheimer, Joe Kennedy, Ro Khanna, Barbara Lee, Zoe Lofgren, Seth Moulton, Frank Pallone, Jackie Speier, and Eric Swalwell. Staff members, including Kenneth DeGraff, Z. J. Hull, Linda Cohen, Thomas Eager, Maher Bitar, Angela Valles, and Slade Bond, have my undying gratitude. Luther Lowe helped me navigate Washington politics. Larry Irving and John Battelle contributed great ideas and helped me take the message to Procter & Gamble.

The former attorney general of New York State, Eric Schneiderman, was the first state AG to take a meeting with me and the first to recognize that internet platforms might be guilty of violating consumer protection laws. My thanks go to Noah Stein and his terrific colleagues.

Thank you to William Schultz and Andrew Goldfarb for great legal advice.

Thank you to Erin McKean for helping me see how the changing perception of Facebook reveals itself in language.

I owe huge thanks to George and Tamiko Soros. George based a speech at the World Economic Forum's Davos conference on my *Washington Monthly* essay and introduced me to an amazing set of ideas and people. Michael Vachon, in particular, has earned my thanks.

One of the most impressive people I met on this journey is Marietje Schaake, a member of the European Parliament from the Netherlands. Marietje is a global thought leader on balancing the needs of society with those of tech platforms. Andrew Rasiej and Micah Sifry of Civic Hall introduced me to Marietje and then guided me through the idealistic and deeply committed world of civic tech. Thank you.

Many people came out of nowhere to help me understand key issues: Ashkan Soltani, Wael Ghonim, Lawrence Lessig, Laurence Tribe, Larry Kramer, Michael Hawley, Jon Vein, Dr. Robert Lustig, Scott Galloway, Chris Hughes, Laura Rosenberger, Karen Kornbluh, Sally Hubbard, T Bone Burnett, Callie Khouri, Daniel Jones, Glenn Simpson, Justin Hendrix, Ryan Goodman, Siva Vaidhyanathan, B. J. Fogg, and Rob and Michele Reiner.

My deepest appreciation to Marc Benioff for supporting the cause early on. Thank you to Tim Berners-Lee for sharing my essay from *Washington Monthly*. Huge thanks to Gail Barnes for being my eyes and ears on social media. Thank you to Alex Knight, Bobby Goodlatte, David Cardinal, Charles Grinstead, Jon Luini, Michael Tchao, Bill Joy, Bill Atkinson, Garrett Gruener, and Andrew Shapiro for ideas, encouragement, and thought-provoking questions. Thank you to Tim Cook and all of Apple for their commitment to protecting the privacy and freedom of customers.

Bono provided invaluable counsel in the early stages of the work that led to this book.

My thanks go out to Herb Sandler, Angelo Carusone, Melissa Ryan, Rebecca Lenn, Eric Feinberg, Gentry Lane, and Christopher Wylie.

Thank you to the Omidyar Network, the Knight Foundation, the Hewlett Foundation, and the Ford Foundation for their support of the Center for Humane Technology.

Thank you Moonalice—Barry Sless, Pete Sears, John Molo, and sometimes Jason Crosby and Katie Skene—for keeping me sane. Special thanks to our entire crew: Dan English, Jenna Lebowitz, Tim Stiegler, Derek Walls, Arthur Rosato, Patrick Spohrer, Joe Tang, Danny Schow, Chris Shaw, Alexandra Fischer, Nick Cernak, Bob Minkin, Rupert Coles, Jamie Soja, Michael Weinstein, and Gail Barnes. Let me also thank the wonderful fans of Moonalice. You know how important you are. Additional thanks to Lebo, Jay Lane, Melvin Seals, Lester Chambers, Dylan Chambers, Darby Gould, Lesley Grant, RonKat Spearman, James Nash, Greg Loiacono, Pete Lavezzoli, Stu Allen, Jeff Pehrson, Dawn Holliday, Grahame Lesh, Alex Jordan, Elliott Peck, Connor O'Sullivan, Bill and Lori Walton, Bob Weir, Mickey Hart, Jeff Chimenti, Rose Solomon, Scott Guberman, the T Sisters, the Brothers Comatose, and Karin Conn.

Massive appreciation to Rory Lenny, Robin Gascon, Dawn Lafond, Gitte Dunn, Diarmuid Harrington, Peter McQuaid, Todd Shipley, Sixto Mendez, Bob Linzy, Fran Mottie, Nick Meriwether, Tim McQuaid, and Jeff Idelson.

My thanks go out to the Other World team, led by Ann McNamee, Hunter Bell, Jeff Bowen, Rebekah Tisch, and Sir Richard Taylor. Your unflagging support sustained me.

Recording the audiobook of *Zucked* was surprisingly fun, thanks to director Brent Katz and engineer Patrick Fitzgerald. Thank you, gentlemen.

The book tour for *Zucked* began the last week of January 2019 and continued without interruption until the end of June. Over the course of five months, I had more than 150 public events, hundreds of press hits, and several hundred other meetings. None of that would have been possible without the support of Liz Calamari at Penguin, Chris Artis at Shreve-Williams, and Ann McNamee. Gregg Sullivan created a brilliant social marketing campaign and encouraged me to invest in ads in the New York City subways. John Borthwick provided the venue for my first event. Thank you to John Markoff, Elizabeth Dwoskin, Kara Swisher, and Judy Estrin for moderating multiple events. Special thanks to Nick Thompson for interviewing me at South By Southwest (SXSW), to Willow Bay and Ted Habte-Gabr for LiveTalks/LA, and to Chris Anderson for the TED Podcast. Thank you to Edward Felsenthal and Lucas Wittman for a wonderful cover story in *Time* magazine. Thank you to the *Times of London* for publishing an extract and a long interview and to *The New York Times* and *The Guardian* for wonderful reviews. Heather Reisman, Bob Ramsey, and Jim Balsillie provided indispensable help in Canada. Thank you, my friends. Thank you to Bob Garfield for having me on his show *On*

The Media, to Joshua Johnson for including me on *1A*, and to Michael Krasny for the opportunity to be on *Forum*. Thanks also to Sam Harris, Kara Swisher, Justin Long, Jason Calacanis, Nick Bilton, James Altucher, Brad Listi, and all the other podcast hosts who welcomed me on their shows. Lastly, I would like to thank the thousands of people who purchased the hardcover edition of *Zucked*, making it a *New York Times* bestseller.

All these people and many more helped to bring this book to life. If I have forgotten anyone, I apologize. Any flaws in this book are my responsibility.

I want to conclude with special thanks to two heroes of mine: Carole Cadwalladr and Clarence Jones. Carole is the bravest journalist I have ever met. Clarence provided strategic and spiritual advice that made my activism successful. You both inspired me.

APPENDIX 1

*This is the essay I sent to Zuck and Sheryl
prior to the election, in October 2016.*

I am really sad about Facebook.

I got involved with the company more than a decade ago and have taken great pride and joy in the company's success . . . until the past few months. Now I am disappointed. I am embarrassed. I am ashamed.

With more than 1.7 billion members, Facebook is among the most influential businesses in the world. Whether they like it or not—whether Facebook is a technology company or a media company—the company has a huge impact on politics and social welfare. Every decision that management makes can matter to the lives of real people. Management is responsible for every action. Just as they get credit for every success, they need to be held accountable for failures. Recently, Facebook has done some things that are truly horrible and I can no longer excuse its behavior.

Here is a sampling of recent actions that conspired to make me so unhappy with Facebook:

- Facebook offers advertisers the opportunity to target advertising by excluding demographic groups such as Blacks and Muslims. In categories such as residential real estate, it is not legal to discriminate on the basis of race. In politics, such a feature would empower racists.

- A 3rd party used Facebook's public APIs to spy on Black Lives Matter. Management's response was, "hey, they are public APIs; anybody can do that." This response is disingenuous, ignoring the fact that Facebook sets the terms of service for every aspect of its site, including APIs, and could easily change the terms to prohibit spying.

- Facebook's facial recognition software is being used to identify people who have not given permission. Given Facebook's apparent willingness to let 3rd parties use its site to spy, the facial recognition raises Fourth Amendment issues of Orwellian proportions.

- Facebook's algorithms have played a huge role in this election cycle by limiting each member's news feed to "things they like," which effectively prevents people from seeing posts that contradict their preconceptions. Trolls on both sides have exploited this bug to spread untruths and inflame emotions.

- Facebook removed human editors from its trending stories list— one of the few ways consumers on FB could be exposed to new stories—and the immediate effect was an explosion in spam stories.

- Facebook is one of many Silicon Valley companies whose employee base is overwhelmingly male and white. Despite having a woman president, Facebook has not made as much progress on diversity as some of its Silicon Valley peers.

- Facebook has publicly defended a board member who has written and spoken publicly against the 19th Amendment and diversity in the workplace.

Lest you think these mistakes were unconscious, consider the following:

- Facebook censors photos in a manner that could be described as puritanical. Recent censorship examples include the Pulitzer prize winning photo of the napalm girl in Vietnam and just about every photograph of breast feeding. Facebook has a prohibition against photographs of female nipples, but does not censor photographs of male nipples.
- Facebook does not allow medical marijuana dispensaries in California to maintain pages, even when they operate in compliance with state laws.
- Facebook has prohibited some promoted posts related to California's Proposition 64–Adult Use of Marijuana Act—claiming that the ads promote something illegal, when in fact they are political ads for a proposition on this year's ballot.

These examples are not equally significant, but as a list they reveal something important: Facebook is not laissez-faire about what happens on its site. Quite the contrary. The only content that stays up on the site is content consistent with Facebook's policies. For a long time, Facebook's policies did no harm, even when they were hard to justify. That is no longer the case. Facebook is enabling people to do harm. It has the power to stop the harm. What it currently lacks is an incentive to do so.

All of these actions took place against a backdrop of extraordinary growth in Facebook usage, revenues and profits. Facebook is among the best performing stocks on Wall Street and one of the most influential companies in the world, which gives many people an excuse to overlook mistakes and reduces the population of people willing to provide constructive criticism. I have reached out directly to management on some of these issues and to members of the press on others; most of my inquiries were ignored and the few responses were unsatisfactory.

It is with considerable reluctance that I am raising these issues through *Recode*. My goal is to persuade Facebook to acknowledge the mistakes it is making, accept responsibility and implement better business practices. Facebook can be successful without being socially irresponsible. We know this because Facebook has chosen to forego a huge economic opportunity in China out of concern for the welfare of the members it would have in that country.

What will it take for Facebook to change? Facebook's executives are a team. In order to empower executives and encourage constructive dissent, they have a policy of sticking together, no matter what. This kind of teamwork has enabled Facebook to grow rapidly and achieve huge scale, but has an ugly side, too. At Facebook, individuals in the senior management team are not punished for mistakes, even big mistakes. As I was told when something disturbing happened several years ago, accountability at Facebook occurs at the team level. This model has produced amazing results—and made so many things better—but it effectively means there is no accountability unless a problem rises to the level where the board would be willing to replace the entire management team. I cannot imagine a realistic scenario where that would happen. So here we are, with increasingly disturbing revelations on what seems like a weekly basis. I want to find a way to encourage Facebook's management to be more socially responsible.

The importance of Facebook's role in our culture and politics cannot be overstated. On current course and speed, Facebook can do more damage in a few months than we can correct in a year. I am hoping to build a coalition of people who share my concern, are willing to engage with Facebook, help them understand our concerns, and work with them to address them. As individuals, we have no power to change Facebook's behavior, but perhaps we can build a coalition large enough to make a difference.

APPENDIX 2

GEORGE SOROS'S DAVOS REMARKS: "THE CURRENT MOMENT IN HISTORY"

Davos, Switzerland, January 25, 2018

The Current Moment in History

Good evening. It has become something of an annual Davos tradition for me to give an overview of the current state of the world. I was planning half an hour for my remarks and half an hour for questions, but my speech has turned out to be closer to an hour. I attribute this to the severity of the problems confronting us. After I've finished, I'll open it up for your comments and questions. So prepare yourselves.

I find the current moment in history rather painful. Open societies are in crisis, and various forms of dictatorships and mafia states, exemplified by Putin's Russia, are on the rise. In the United States, President Trump would like to establish a mafia state but he can't, because the Constitution, other institutions, and a vibrant civil society won't allow it.

Whether we like it or not, my foundations, most of our grantees and myself personally are fighting an uphill battle, protecting the democratic achievements of the past. My foundations used to focus on the so-called developing world, but now that the open society is also endangered in the United States and Europe, we are spending more than half our budget closer to home because what is happening here is having a negative impact on the whole world.

But protecting the democratic achievements of the past is not enough; we must also safeguard the values of open society so that they will better withstand future onslaughts. Open society will always have its enemies, and each generation has to reaffirm its commitment to open society for it to survive.

The best defense is a principled counterattack. The enemies of open society feel victorious and this induces them to push their repressive efforts too far, this generates resentment and offers opportunities to push back. That is what is happening in places like Hungary today.

I used to define the goals of my foundations as "defending open societies from their enemies, making governments accountable and fostering a critical mode of thinking." But the situation has deteriorated. Not only the survival of open society, but the survival of our entire civilization is at stake. The rise of leaders such as Kim Jong-Un in North Korea and Donald Trump in the US have much to do with this. Both seem willing to risk a nuclear war in order to keep themselves in power. But the root cause goes even deeper.

Mankind's ability to harness the forces of nature, both for constructive *and* destructive purposes, continues to grow while our ability to govern ourselves properly fluctuates, and it is now at a low ebb.

The threat of nuclear war is so horrendous that we are inclined to ignore it. But it is real. Indeed, the United States is set on a course toward nuclear war by refusing to accept that North Korea has become a nuclear power. This creates a strong incentive for North Korea to develop its nuclear capacity with all possible speed, which in turn may induce the United States to use its nuclear superiority preemptively; in effect to start a nuclear war in order to prevent nuclear war—an obviously self-contradictory strategy.

The fact is, North Korea has become a nuclear power and there is no military action that can prevent what has already happened. The only sensible strategy is to accept reality, however unpleasant it is, and to come to terms with North Korea as a nuclear power. This requires the United States to cooperate with all the interested parties, China foremost among them. Beijing holds most of the levers of power against North Korea, but is reluctant to use them. If it came down on Pyongyang too hard, the regime could collapse and China would be flooded by North Korean refugees. What is more, Beijing is reluctant to do any favors for the United States, South Korea or Japan—against each of which it harbors a variety of grudges. Achieving cooperation will require extensive negotiations, but once it is attained, the alliance would be able to confront North Korea with both carrots and sticks. The sticks could be used to force it to enter into good faith negotiations and the carrots to reward it for verifiably suspending further development of nuclear weapons. The sooner a so-called freeze-for-freeze agreement can be reached, the more successful the policy will be. Success can be measured by the amount of time it would take for North Korea to make its nuclear arsenal fully operational. I'd like to draw your attention to two seminal reports just published by Crisis Group on the prospects of nuclear war in North Korea.

The other major threat to the survival of our civilization is climate change, which is also a growing cause of forced migration. I have dealt with the problems of migration at great length elsewhere, but I must emphasize how severe and intractable those problems are. I don't want to go into details on climate change either because it is well known what needs to be done. We have the scientific knowledge; it is the political will that is missing, particularly in the Trump administration.

Clearly, I consider the Trump administration a danger to the world. But I regard it as a purely temporary phenomenon that will disappear in 2020, or even sooner. I give President Trump credit for motivating his core supporters brilliantly, but for every core supporter, he has created a greater number of core opponents who are equally strongly motivated. That is why I expect a Democratic landslide in 2018.

My personal goal in the United States is to help reestablish a functioning two-party system. This will require not only a landslide in 2018 but also a Democratic Party that will aim at non-partisan redistricting, the appointment of well-qualified judges, a properly conducted census and other measures that a functioning two-party system requires.

The IT Monopolies

I want to spend the bulk of my remaining time on another global problem: the rise and monopolistic behavior of the giant IT platform companies. These companies have often played an innovative and liberating role. But as Facebook and Google have grown into ever more powerful monopolies, they have become obstacles to innovation, and they have caused a variety of problems of which we are only now beginning to become aware.

Companies earn their profits by exploiting their environment. Mining and oil companies exploit the physical environment; social media companies exploit the social environment. This is particularly nefarious because social media companies influence how people think and behave without them even being aware of it. This has far-reaching adverse consequences on the functioning of democracy, particularly on the integrity of elections.

The distinguishing feature of internet platform companies is that they are networks and they enjoy rising marginal returns; that accounts for their phenomenal growth. The network effect is truly unprecedented and transformative, but it is also unsustainable. It took Facebook eight and a half years to reach a billion users and half that time to reach the second billion. At this rate, Facebook will run out of people to convert in less than three years.

Facebook and Google effectively control over half of all internet advertising revenue. To maintain their dominance, they need to expand their networks *and* increase their share of users' attention. Currently they do this by providing users with a convenient platform. The more time users spend on the platform, the more valuable they become to the companies.

Content providers also contribute to the profitability of social media companies because they cannot avoid using the platforms and they have to accept whatever terms they are offered.

The exceptional profitability of these companies is largely a function of their avoiding responsibility for—and avoiding paying for—the content on their platforms.

They claim they are merely distributing information. But the fact that they are near-monopoly distributors makes them public utilities and should subject them to more stringent regulations, aimed at preserving competition, innovation, and fair and open universal access.

The business model of social media companies is based on advertising. Their true customers are the advertisers. But gradually a new business model is emerging, based not only on advertising but on selling products and services directly to users. They exploit the data they control, bundle the services they offer and use discriminatory pricing to keep for themselves more of the benefits that otherwise they would have to share with consumers. This enhances their profitability even further—but the bundling of services and discriminatory pricing undermine the efficiency of the market economy.

Social media companies deceive their users by manipulating their attention and directing it towards *their* own commercial purposes. They deliberately engineer addiction to the services they provide. This can be very harmful, particularly for adolescents. There is a similarity between internet platforms and gambling companies. Casinos have developed techniques to hook gamblers to the point where they gamble away all their money, even money they don't have.

Something very harmful and maybe irreversible is happening to human attention in our digital age. Not just distraction or addiction; social media companies are inducing people to give up their autonomy. The power to shape people's attention is increasingly concentrated in the hands of a few companies. It takes a real effort to assert and defend what John Stuart Mill called "the freedom of mind." There is a possibility that once lost, people who grow up in the digital age will have difficulty in regaining it. This may have far-reaching political consequences. People without the freedom of mind can be easily manipulated. This danger does not loom only in the future; it already played an important role in the 2016 US presidential election.

But there is an even more alarming prospect on the horizon. There could be an alliance between authoritarian states and these large, data-rich IT monopolies that would bring together nascent systems of corporate surveillance with an already developed system of state-sponsored surveillance. This may well result in a web of totalitarian control the likes of which not even Aldous Huxley or George Orwell could have imagined.

The countries in which such unholy marriages are likely to occur first are Russia and China. The Chinese IT companies in particular are fully equal to the American ones. They also enjoy the full support and protection of the Xi Jingping regime. The government of China is strong enough to protect its national champions, at least within its borders.

US-based IT monopolies are already tempted to compromise themselves in order to gain entrance to these vast and fast-growing markets. The dictatorial leaders in these countries may be only too happy to collaborate with them since they want to improve their methods of control over their own populations and expand their power and influence in the United States and the rest of the world.

The owners of the platform giants consider themselves the masters of the universe, but in fact they are slaves to preserving their dominant position. It is only a matter of time before the global dominance of the US IT monopolies is broken. Davos is a good place to announce that their days are numbered. Regulation and taxation will be their undoing and EU Competition Commissioner Vestager will be their nemesis.

There is also a growing recognition of a connection between the dominance of the platform monopolies and the rising level of inequality. The concentration of share ownership in the hands of a few private individuals plays some role but the peculiar position occupied by the IT giants is even more important. They have achieved monopoly power, but at the same time they are also competing against each other. They are big enough to swallow startups that could develop into competitors, but only the giants have the resources to invade each other's territory. They are poised to dominate the new growth areas that artificial intelligence is opening up, like driverless cars.

The impact of innovations on unemployment depends on government policies. The European Union and particularly the Nordic countries are much more farsighted in their social policies than the United States. They protect the workers, not the jobs. They are willing to pay for re-training or retiring displaced workers. This gives workers in Nordic countries a greater sense of security and makes them more supportive of technological innovations than workers in the US.

The internet monopolies have neither the will nor the inclination to protect society against the consequences of their actions. That turns them into a menace, and it falls to the regulatory authorities to protect society against them. In the United States, the regulators are not strong enough to stand up against their political influence. The European Union is better situated because it doesn't have any platform giants of its own.

The EU uses a different definition of monopoly power from the United States. US law enforcement focuses primarily on monopolies created by acquisitions, whereas EU law prohibits the abuse of monopoly power irrespective of how it is achieved. Europe has much stronger privacy and data protection laws than America. Moreover, US law has adopted a strange doctrine: it measures harm as an increase in the price paid by customers for services received—and that is almost impossible to prove when most services are provided for free. This leaves out of consideration the valuable data platform companies collect from their users.

Commissioner Vestager is the champion of the European approach. It took the EU seven years to build a case against Google, but as a result of her success the process has been greatly accelerated. Due to her proselytizing, the European approach has begun to affect attitudes in the United States as well.

The Rise of Nationalism and How to Reverse It

I have mentioned several of the most pressing and important problems confronting us today. In conclusion, let me point out that we are living in a revolutionary period. All our established institutions are in a state of flux and in these circumstances both fallibility and reflexivity are operating at full force.

I lived through similar conditions in my life, most recently some thirty years ago. That's when I set up my network of foundations in the former Soviet empire. The main difference between the two periods is that thirty years ago the dominant creed was international governance and cooperation. The European Union was the rising power and the Soviet Union the declining one. Today, however, the motivating force is nationalism. Russia is resurgent and the European Union is in danger of abandoning its values.

As you will recall, the previous experience didn't turn out well for the Soviet Union. The Soviet empire collapsed and Russia has become a mafia state that has adopted a nationalist ideology. My foundations did quite well: the more advanced members of the Soviet empire joined the European Union.

Now our aim is to help save the European Union in order to radically reinvent it. The EU used to enjoy the enthusiastic support of the people of my generation, but that changed after the financial crisis of 2008. The EU lost its way because it was governed by outdated treaties and a mistaken belief in austerity policies. What had been a voluntary association of equal states was converted into a relationship between creditors and debtors where the debtors couldn't meet their obligations and the creditors set the conditions that the debtors had to meet. That association was neither voluntary nor equal.

As a consequence, a large proportion of the current generation has come to regard the European Union as its enemy. One important country, Britain, is in the process of leaving the EU and at least two countries, Poland and Hungary, are ruled by governments that are adamantly opposed to the values on which the European Union is based. They are in acute conflict with various European institutions and those institutions are trying to discipline them. In several other countries anti-European parties are on the rise. In Austria, they are in the governing coalition, and the fate of Italy will be decided by the elections in March.

How can we prevent the European Union from abandoning its values? We need to reform it at every level: at the level of the Union itself, at the level of the member states and the level of the electorate. We are in a revolutionary period; everything is subject to change. The decisions taken now will determine the shape of the future.

At the Union level, the main question is what to do about the euro. Should every member state be required to eventually adopt the euro, or should the current situation be allowed to continue indefinitely? The Maastricht Treaty prescribed the first alternative but the euro has developed some defects that the Maastricht Treaty didn't foresee and still await resolution.

Should the problems of the euro be allowed to endanger the future of the European Union? I would strongly argue against it. The fact is that the countries that don't qualify are eager to join the euro, but those that do have decided against it, with the exception of Bulgaria. In addition, I should like to see Britain remain a member of the EU or eventually rejoin it, and that couldn't happen if it meant adopting the euro.

The choice confronting the EU could be better formulated as one between a multispeed and a multitrack approach. In a multispeed approach, member states have to agree in advance on the ultimate outcome; in a multitrack approach, member states are free to form coalitions of the willing to pursue particular goals on which they agree. The multitrack approach is obviously more flexible but the European bureaucracy favored the multispeed approach. That was an important contributor to the rigidity of the EU's structure.

At the level of the member states, their political parties are largely outdated. The old distinction between left and right is overshadowed by being either pro- or anti-European. This manifests itself differently in different countries.

In Germany, the Siamese twin arrangement between the CDU and the CSU has been rendered unsustainable by the results of the recent elections. There is another party, the AfD further to the right than the CSU in Bavaria. This has forced the CSU to move further to the right in anticipation of next year's local elections in Bavaria so that the gap between the CSU and the CDU has become too great. This has rendered the German party system largely dysfunctional until the CDU and CSU break up.

In Britain, the Conservatives are clearly the party of the right and Labor the party of the left, but each party is internally divided in its attitude toward Brexit. This complicates the Brexit negotiations immensely, and makes it extremely difficult for Britain as a country to decide and modify its position towards Europe.

Other European countries can be expected to undergo similar

realignments with the exception of France, which has already undergone its internal revolution.

At the level of the electorate, the top-down initiative started by a small group of visionaries led by Jean Monnet carried the process of integration a long way, but it has lost its momentum. Now we need a combination of the top-down approach of the European authorities with the bottom-up initiatives started by an engaged electorate. Fortunately, there are many such bottom-up initiatives; it remains to be seen how the authorities will respond to them. So far, President Macron has shown himself most responsive. He campaigned for the French presidency on a pro-European platform, and his current strategy focuses on the elections for the European Parliament in 2019, and that requires engaging the electorate.

While I have analyzed Europe in greater detail, from a historical perspective what happens in Asia is ultimately much more important. China is the rising power. There were many fervent believers in the open society in China who were sent to be reeducated in rural areas during Mao's Revolution. Those who survived returned to occupy positions of power in the government. So the future direction of China used to be open-ended, but no more.

The promoters of open society have reached retirement age, and Xi Jinping, who has more in common with Putin than with the so-called West, has begun to establish a new system of party patronage. I'm afraid that the outlook for the next twenty years is rather bleak. Nevertheless, it's important to embed China in institutions of global governance. This may help to avoid a world war that would destroy our entire civilization.

That leaves the local battlegrounds in Africa, the Middle East, and Central Asia. My foundations are actively engaged in all of them. We are particularly focused on Africa, where would-be dictators in Kenya, Zimbabwe, and the Democratic Republic of Congo have perpetrated electoral fraud on an unprecedented scale, and citizens are literally risking their lives to resist the slide into dictatorship. Our goal is to empower local people to deal with their own problems, assist the disadvantaged and reduce human suffering to the greatest extent possible. This will leave us plenty to do well beyond my lifetime.

BIBLIOGRAPHIC ESSAY

Normally I prefer to read novels. In the three years since I first realized there was a problem at Facebook, I have read several novels and many nonfiction volumes that helped me understand that problem. In this essay, I want to share my intellectual journey but also point to books and other media that shed light on the people, business practices, and culture that enabled it.

My education about the dark side of social media began in 2011 with Eli Pariser's groundbreaking TED Talk on filter bubbles. I recommend the video of that talk, as well as Eli's book, *The Filter Bubble: What the Internet Is Hiding from You* (New York: Penguin Press, 2012).

The book that energized me in early 2017 was *Move Fast and Break Things: How Facebook, Google, and Amazon Cornered Culture and Undermined Democracy*, by Jonathan Taplin (New York: Little, Brown and Company, 2017). Taplin had brilliant careers in rock 'n' roll and Hollywood before moving to academia. He describes how internet platforms exploited a legal safe harbor to take over American culture, in the process causing serious damage to democracy. Tim Wu's *The Attention Merchants: The Epic Scramble to Get Inside Our Heads* (New York: Alfred A. Knopf, 2016) explains the history of persuasion for profit from tabloid newspapers through social media. It is essential reading. *World Without Mind: The Existential Threat of Big Tech*, by Franklin Foer (New York: Penguin Press, 2017), sharpens the argument against unconstrained capitalism on the internet. Both Taplin and Foer pine for the past in media, when everything was more genteel and the public square remained vibrant, but that does not prevent them from contributing to the critical discussion about what needs to be done.

A very early warning about the dangers of technology for democracy came from Evgeny Morozov in his book *The Net Delusion: The Dark Side of Internet Freedom* (New York: PublicAffairs, 2011). Morozov described how the same tools that empower democratic movements can be co-opted by authoritarians to suppress democracy. Morozov's second book, *To Save Everything, Click Here: The Folly of Technological Solutionism* (New York: PublicAffairs, 2013), addresses the implications of Big Data for society.

Relative to Facebook, the place to start is with *The Facebook Effect: The Inside Story of the Company That Is Connecting the World*, by David Kirkpatrick (New York: Simon & Schuster, 2010). Written early in Facebook's history, this book exudes the optimism that I and so many others felt about Facebook's future, but Kirkpatrick also manages to foreshadow almost every problem triggered by Facebook's success. As a companion, I recommend *Antisocial Media: How Facebook Disconnects Us and Undermines Democracy*, by Siva Vaidhyanathan (New York: Oxford University Press, 2018). *Antisocial Media* pulls back the curtain on Facebook's technology and what it really does to users. This is a must-read.

Zuck and others criticize the film *The Social Network* and the book on which it was based, *The Accidental Billionaires: The Founding of Facebook: A Tale of Sex, Money, Genius, and Betrayal*, by Ben Mezrich (New York: Anchor, Reprint edition 2010), as inaccurate to the point of being fiction. The sad thing is how consistent the personalities and behaviors depicted in these stories are with the Facebook people caught in the glare of election interference and Cambridge Analytica.

There are several good books about the culture in Silicon Valley. A good place to start is *Brotopia: Breaking Up the Boys' Club of Silicon Valley*, by Emily Chang (New York: Portfolio, 2018). Chang is the host of *Bloomberg Technology*, the show on which I interviewed Tristan in April 2017, after his appearance on *60 Minutes*. (Emily was on maternity leave that day!) The domination of Silicon Valley by young Asian and Caucasian men seems foundational to the culture that built Facebook, YouTube, and the others. Chang cuts to the heart of the matter.

Chaos Monkeys: Obscene Fortune and Random Failure in Silicon Valley, by Antonio García Martínez (New York: Harper, 2016), is the inside story of an engineer who started a company, ran out of money, got a gig in advertising technology at Facebook, and was there during the formative years of the business practices that enabled Cambridge Analytica. Through the lens of this book, you will get a clear view of the culture and internal practices of Facebook and other platforms.

Valley of Genius: The Uncensored History of Silicon Valley, by Adam Fisher (New York: Twelve, 2018), is a collection of interviews and quotes from participants at nearly every stage of Silicon Valley's development. The book begins in the sixties, with Douglas Engelbart, and marches steadily to the present. The interview with Facebook founders and early employees is essential reading. There are also illuminating interviews with the early people at Google and Twitter.

There are two works of fiction that prepared me to recognize the disturbing signals I picked up from Facebook in 2016. *The Circle*, by Dave Eggers (New York: Alfred A. Knopf, 2013), describes a fictional company that combines the attributes of Facebook and Google. Written as science fiction, the issues it raises came to market only a few years after publication. *After On*, by Rob Reid (New York: Del Rey, 2017), imagines a next-generation social network whose AI becomes sentient. This is a long but incredibly funny novel that is worth every moment you spend with it. You will understand why technologists must be forced to prepare for unintended consequences. They always happen, and their impact is increasingly harmful.

HBO's television series *Silicon Valley* lampoons the startup culture in a way that always rings true. The plots are exaggerated, but not by as much as you might think. The Valley culture is strange. When I first met the creative team, I asked the boss man, Mike Judge, to describe the gestalt of the show. He said, "There is a titanic struggle between the hippie culture espoused by people like Steve Jobs and the libertarian culture of Peter Thiel." I responded, "And the libertarians are winning." He smiled. I looked around the table and realized why I was there. "Because I'm one of the last hippies." They all smiled. Full disclosure: I have been a technical consultant on *Silicon Valley* since season 2.

After I met Tristan Harris, I spent several months learning everything I could about persuasive technology, from the underlying psychology to the ways it can be implemented in software. I started with Tristan's many interviews and blog posts before reading the textbook *Persuasive Technology: Using Computers to Change What We Think and Do*, by B. J. Fogg (San Francisco: Morgan Kaufmann, 2002), the Stanford University professor who taught Tristan and so many others. The textbook is well written, which made me realize how easy it would have been for students to grasp the techniques of persuasive technology and appreciate its power. The layman's version of Fogg's book is *Hooked: How to Build Habit-Forming Products*, by Nir Eyal (New York: Portfolio, 2016). Eyal is an unapologetic advocate for persuasive technology, an addiction-denier who has built a successful business teaching entrepreneurs and engineers how to exploit human psychology in software.

Once I understood how persuasive technology works, I consulted two fantastic books on the psychological impact of persuasive technology on smartphones, tablets, and computers. *Irresistible: The Rise of Addictive Technology and the Business of Keeping Us Hooked*, by Adam Alter (New York: Penguin Press, 2017), is comprehensive, well written, and easy to understand. It covers a wide range of harms across every age group. A must read. *Glow Kids: How Screen Addiction Is Hijacking Our Kids—and How to Break the Trance*, by Nicholas Kardaras (New York: St. Martin's Press, 2016), focuses on kids. If you have young children, I predict this book will cause you to limit their exposure to screens and to protect them from a range of applications. Another important book is *It's Complicated: The Social Lives of Networked Teens*, by danah boyd (New Haven: Yale University Press, 2014). This should be must reading for parents.

Social media has taken over the public square in almost every country where Facebook and other platforms operate. *Twitter and Tear Gas: The Power and Fragility of Networked Protest*, by Zeynep Tufekci (New Haven: Yale University Press, 2017), probes the use of social media by protesters and the counterattacks of the powerful. Tufekci writes a monthly opinion piece in *The New*

York Times that smartens me up every time. This book helped me understand what really happened in the Arab Spring and why the internet platforms' current incentives always favor those in power. *Messing with the Enemy: Surviving in a Social Media World of Hackers, Terrorists, Russians, and Fake News*, by Clint Watts (New York: Harper, 2018), is the best book I found on the use of social media by bad actors, and it's great! Clint was an agent at the FBI who now spends his time monitoring hostile activity on social media. I have met Clint a few times around MSNBC; his insights profoundly improved this book. *The Square and the Tower: Networks and Power, from the Freemasons to Facebook*, by Niall Ferguson (New York: Penguin Press, 2018), puts the power of Facebook and Google into historical context. Scary.

Ten Arguments for Deleting Your Social Media Accounts Right Now, by Jaron Lanier (New York: Henry Holt and Co., 2018), is short and sweet. Lanier has been a thought leader in technology for three decades, with an early emphasis on virtual reality, but in this book he speaks as a concerned technologist who is also a philosopher about technology. This book did not need ten arguments, but I learned something from every one. One of Lanier's major concerns—unrestricted development of artificial intelligence—is the subject of *Machines of Loving Grace*, by John Markoff (New York: Ecco, 2015). The book explains how artificial intelligence risks undermining humans, rather than leveraging them.

To understand the world of Big Data, and the challenges it poses to society, I recommend Cathy O'Neil's *Weapons of Math Destruction* (New York: Crown, 2016), which has the best explanation of the good, the bad, and the ugly of algorithms that I have ever read. Yuval Harari's *21 Lessons for the 21st Century* (New York: Spiegel & Grau, 2018), explores the implication of a future where robots and artificial intelligence threaten the relevance of humans in the economy.

It is hard to appreciate fully the threat from internet platforms without understanding the business philosophy that gave rise to them. *The Lean Startup: How Today's Entrepreneurs Use Continuous Innovation to Create Radically Successful Businesses*, by Eric Ries (New York: Currency, 2011), was the bible for entrepreneurs and venture capitalists from the time Ries first proposed it in 2008. *Zero to One: Notes on Startups, or How to Build the Future*, by Peter Thiel (New York: Crown Business, 2014), is a glimpse inside the mind of Facebook's first outside investor. Libertarianism remains fashionable in Silicon Valley, though not everyone has embraced Thiel's version of it. *The Four: The Hidden DNA of Amazon, Apple, Facebook, and Google*, by Scott Galloway (New York: Portfolio, 2017), is an unusually insightful business book. The former entrepreneur and current professor at the NYU Stern School of Business combines a deep understanding of economics and business with trenchant analyses of management and strategy.

The months after I finished writing *Zucked* saw the release of four profoundly valuable books. No book had a greater impact on my journey of discovery than *The Age of Surveillance Capitalism: The Fight for a Human Future at the New Frontier of Power* (New York: PublicAffairs, 2019), by Shoshana Zuboff. It details the most important economic model of our times, naming the key elements, explaining how surveillance capitalism developed and how it is likely to evolve. If you only read one other book about this topic, Zuboff's should be the one. The fact that "surveillance capitalism" is now in common parlance is evidence of this book's impact. Many will be intimidated by the size of the book—535 pages of text—but I found value on every page. This book should be required reading for every policy maker and concerned citizen. If you prefer interviews and podcasts, there are many great ones with Shoshana Zuboff. One of my favorites is RecodeDecode with Kara Swisher.

The Road to Unfreedom: Russia, Europe, America (New York: Tim Duggan Books, 2018), by Timothy Snyder tracks the evolution of Russia's use of social media to disrupt its enemies. After studying the return of totalitarian thought to Russia in 2011 and the death of democracy in 2012 as they happened, Snyder followed the story as Russia deployed psychological and information operations against members of the European Union and the United States. This book exposes the fragility of western democratic institutions in the face of a well-organized campaign of disinformation and hate speech. Snyder makes a compelling case that democracy begins with truth, which enables trust and individuality. Conformity and complacency are the enemies of democracy and freedom.

Future Politics: Living Together in a World Transformed by Tech (Oxford: Oxford University Press, 2018), by Jamie Susskind proposes a new political theory for a world where code and algorithms have more impact on daily life than the law. *Like War: The Weaponization of Social Media* (Boston: Houghton Mifflin Harcourt, 2018), by P. W. Singer and Emerson T. Brooking is a deep dive into the use of internet platforms as an alternative to armaments as a tool of politics and foreign policy.

Other recent books of note: *Network Propaganda: Manipulation, Disinformation, and Radicalization in American Politics* (New York: Oxford University Press, 2018), by Yochai Benkler, Robert Faris, and Hal Roberts, academics at the Berkman Klein Center for Internet and Society at Harvard. Oxford professor Rachel Botsman wrote *Who Can You Trust?: How Technology Brought Us Together and Why It Might Drive Us Apart* (New York: PublicAffairs, 2018) to help readers fight back against the divisive power of internet platforms. *The Code: Silicon Valley and the Remaking of America* (New York: Penguin Press, 2019), by Margaret O'Mara traces the development of the technology industry, with a focus on its partnership with and support from the US government. *The Right of Publicity: Privacy Reimagined for a Public World* (Cambridge: Harvard University Press, 2018), by Jennifer E. Rothman transforms the conversation about privacy by applying a legal principle traditionally employed only by celebrities to all citizens.

FEW PEOPLE REALIZE HOW much Silicon Valley has changed since 2000. Culturally, there is no comparison to the eighties or nineties. I think of Silicon Valley's culture as having three distinct eras: Apollo, Hippie, and Libertarian. In the Apollo era, the engineers were white males in short-sleeve white shirts, with a tie and plastic pocket protector. The hippie era, which began with Atari, soon to be followed by Apple, created video games and personal computers. It faded in the nineties, before dying out when the internet bubble burst in 2000. The first two eras saw great inventions and huge growth, creating a giant reservoir of goodwill and trust with consumers and policy makers. The libertarian era that began just after 2000 transformed the value system, culture, and business models of Silicon Valley, enabling Google, Facebook, and Amazon to create unprecedented wealth by dominating the public square on a global basis. I liked the Apollo and hippie eras because their idealism was genuine, even when it was misdirected. There are some excellent books that will unlock the mysteries of Silicon Valley's early and middle days. Please read them!

A good place to start is with semiconductors, the original business of Silicon Valley. I recommend *The Man Behind the Microchip: Robert Noyce and the Invention of Silicon Valley*, by Leslie Berlin (New York: Oxford University Press, 2005). Noyce is a thread from the transistor to the Intel microprocessor. *The Dream Machine: J. C. R. Licklider and the Revolution That Made Computing Personal*, by M. Mitchell Waldrop (New York: Viking, 2001), explains how the idea of personal computing came to be. *Bootstrapping: Douglas Engelbart, Coevolution, and the Origins of Personal Computing*, by Thierry Bardini (Palo Alto: Stanford University Press, 2000), tells the story of the genius who created the mouse, visualized a networked world of PCs, and gave the Mother of All Demos. *Dealers of Lightning: Xerox PARC and the Dawn of the Computer Age*, by Michael A. Hiltzik (New York: HarperBusiness, 1999), takes the reader inside the research center in Palo Alto where Steve Jobs saw the future. *Troublemakers: Silicon Valley's Coming of Age*, by Leslie Berlin (New York: Simon & Schuster, 2017), tells the story of the men and women, some well known, others obscure, who helped to build Silicon Valley. *The Innovators: How a Group of Hackers, Geniuses and Geeks Created the Digital Revolution*, by Walter Isaacson (New York: Simon & Schuster, 2014), looks back on the key people whose work created Silicon Valley. *What the Dormouse Said: How the 60s Counterculture Shaped the Personal Computer Industry*, by John Markoff (New York: Viking, 2005), shows how hippie culture became the culture of the PC industry. Tom Wolfe's *The Electric Kool-Aid Acid Test* (New York: Farrar Straus and Giroux, 1968) is a helpful introduction to the culture embraced by a core group in Silicon Valley at a critical time. *Fire in the Valley: The Making of the Personal Computer*, by Paul Freiberger and Michael Swaine (Berkeley: Osborne/McGraw-Hill, 1984), is the best book I know on the early days of the personal computer industry, from computer clubs to the start of Microsoft and Apple, to the battle that followed. The revised edition, which bears a different subtitle, follows the industry into its declining years. *Hackers: Heroes of the Computer Revolution*, by Steven Levy (New York: Anchor Press/Doubleday, 1984), investigates a key subculture in Silicon Valley. Levy wrote this as it was happening, which makes the book particularly helpful, as in the case of *The Facebook Effect*. *Cyberpunk: Outlaws and Hackers on the Computer Frontier*, by Katie Hafner and John Markoff (New York: Simon & Schuster, 1991), picks up the story of *Hackers* and carries it forward.

I RECOMMEND LEARNING ABOUT the origin stories of the other internet platforms. *The Everything Store: Jeff Bezos and the Age of Amazon*, by Brad Stone (New York: Little, Brown and Co., 2013), blew my mind. I remember the day Jeff Bezos first presented to my partners, the venture capital firm of Kleiner Perkins Caufield & Byers. There is a very strong argument that the success of Amazon represents the greatest accomplishment of any startup since 1990. Bezos is amazing. His relatively low profile masks the pervasive influence of his company.

Like *The Facebook Effect*, *In the Plex: How Google Thinks, Works, and Shapes Our Lives*, by Steven Levy (New York: Simon & Schuster, 2011), is exceptionally sympathetic to its subject. That is the price of getting access to a tech giant. As long as you remind yourself that Google adjusts search results based on what it perceives to be your interests and that YouTube's algorithms promote conspiracy theories, you will get a great deal of value from this book. *Hatching Twitter: A True Story of Money, Power, Friendship, and Betrayal*, by Nick Bilton (New York, Portfolio, 2013), is worth reading because of Twitter's outsized influence on journalists, an influence out of proportion with the skills of Twitter's leadership.

There were many books written about Microsoft in the glory days of the 1990s. To understand the early years, I would recommend *Hard Drive: Bill Gates and the Making of the Microsoft Empire* (New York: HarperBusiness,

1993) by James Wallace and Jim Erickson, and *Gates: How Microsoft's Mogul Reinvented an Industry—and Made Himself the Richest Man in America* (New York: Simon and Schuster, 1994), by Stephen Manes and Paul Andrews. Recently, the top Microsoft executives have written books to celebrate the company's extraordinary return to industry leadership and high growth. CEO Satya Nadella wrote *Hit Refresh: The Quest to Rediscover Microsoft's Soul and Imagine a Better Future for Everyone* (New York: HarperBusiness, 2017). The company's president, Brad Smith, and Carol Ann Browne just put out *Tools and Weapons: The Promise and The Peril of the Digital Age* (New York: Penguin Press, 2019). When reading these books, keep in mind Microsoft's increasing commitment to and success in surveillance capitalism.

No study of Silicon Valley would be complete without a focus on Steve Jobs. Walter Isaacson's biography *Steve Jobs* (New York: Simon & Schuster, 2011) was a bestseller. I was lucky enough to know Steve Jobs. We were not close, but I knew Steve for a long time and had several opportunities to work with him. I experienced the best and the worst. Above all, I respect beyond measure all the amazing products created on Steve's watch.

This bibliographic essay includes only the books that helped me prepare to write *Zucked*. There are other fine books on these topics.

INDEX